PRAISE FOR THE USE OF MEDIEVAL WEAPONS

Eric's book is the best one-stop survey of the wide swath of medieval European martial arts published so far. Part history lesson and part primer for enthusiasts, Eric's book is very much written in the voice of a seasoned enthusiast passing down what seems to be unrecorded wisdom to a new generation. He successfully juggles the real-world fighting guidance of over a dozen historical masters with the kind of joy that all modern swordspeople experience, best captured in his mantra, "We're all sword geeks here."

—Jake Norwood, President of Longpoint HEMA Tournaments and Workshops, founding member of the HEMA Alliance, and sword geek

Eric Lowe's *Use of Medieval Weaponry* is an introduction to Historical European Martial Arts takes a novel approach: he breaks down his survey not by the various fighting tradition of the Middle Ages and Renaissance, but by instead examining the different types of weapons in use; and only then addressing how each martial school employed and taught their use. The result is a whirlwind tour of several centuries of martial arts, supplemented with some very entertaining thoughts and

speculations on how these weapons would fare when pitted against each other. The book is sure to appeal to today's historians, martial practitioners, and combat researchers.
—Christian Henry Tobler, author of
Fighting with the German Longsword

Lowe's survey of medieval and Early Renaissance fencing distills decades of HEMA research into a delightful and informative read. This one-of-a-kind book explains how people fought five hundred years ago and why they fought that way. This is a serious book that does not take itself too seriously—a perfect read for those new to our peculiar hobby or anyone seeking to conjure up the dusty battlefields and bloody streets of yore.
—Stephen Fratus, Author of With *Malice & Cunning: Anonymous 16th Century Manuscript on Bolognese Swordsmanship*

This book is a wonderful introduction to some of the most popular European weapons of the medieval and Renaissance periods. The author shows a great understanding of their use, and informs the reader in a very enjoyable fashion. There is something to learn for those who have limited knowledge to those with a more extensive one. If you are looking for a book to help you get your toes wet in either European weaponry and their usage, or Historical European Martial Arts in general, this is a great place to start.
—Keith Cotter-Reilly, Head Instructor,
Atlanta Historical Fencing Academy

A great bird's-eye view of trends in how people fought in the medieval and early Modern periods. This book covers a good variety of the most popular weapons of the period, including various swords, polearms, and secondary weapons, and explains in some detail what historical treatises have to say about them. If you're looking for an introduction to the subject of medieval martial arts, Eric Lowe has you covered.
—Michael Chidester, Wiktenauer Director

THE USE OF
MEDIEVAL WEAPONRY

THE USE OF MEDIEVAL WEAPONRY

Eric Lowe

AEON

First published in 2020 by
Aeon Books Ltd
12 New College Parade
Finchley Road
London NW3 5EP

Copyright © 2020 by Eric Lowe

The right of Eric Lowe to be identified as the author of this work has been asserted in accordance with §§ 77 and 78 of the Copyright Design and Patents Act 1988.

All rights reserved. No part of this publication may be reproduced, stored in a retrieval system, or transmitted, in any form or by any means, electronic, mechanical, photocopying, recording, or otherwise, without the prior written permission of the publisher.

British Library Cataloguing in Publication Data

A C.I.P. for this book is available from the British Library

ISBN-13: 978-1-91280-726-0

Typeset by Medlar Publishing Solutions Pvt Ltd, India
Printed in Great Britain

www.aeonbooks.co.uk

CONTENTS

ACKNOWLEDGMENTS — ix

A NOTE ON TRANSLATIONS — xi

PREFACE — xv

CHAPTER ONE
We're all sword geeks here — 1

CHAPTER TWO
How do we know any of this? — 5

CHAPTER THREE
Of royal blood: longsword — 17

CHAPTER FOUR
Excellent and useful: sword and buckler — 41

CHAPTER FIVE
Queen of swords: sword and shield — 71

CHAPTER SIX
The forgotten: sword and cloak, sword and dagger,
  and two swords   97

CHAPTER SEVEN
Malice and art: knives and daggers   131

CHAPTER EIGHT
Many obligations: greatswords   147

CHAPTER NINE
The unseen: what about axes and maces?   171

CHAPTER TEN
Arm's length: thrusting polearms   179

CHAPTER ELEVEN
Grace and results: cutting polearms   203

CHAPTER TWELVE
Taking hits: fencing in, and against, armor   219

CHAPTER THIRTEEN
Honor in wars: military fencing   243

CHAPTER FOURTEEN
Final thoughts   257

WORKS CITED   259

INDEX   263

ACKNOWLEDGMENTS

I am deeply grateful to everybody whose actions, however small, have conspired to put a sword in my hand. You have all made my teenaged self's dream come truer than I ever thought possible. In particular, my thanks go out to:

Tristan Zukowski, Michael Edelson, and Sang Kim, for your teaching;

My own students at Swordwind Historical Swordsmanship, past, present, and future, for challenging me to understand our art more deeply, for giving me the opportunity to pursue it more seriously, and for your own dedication;

Mary and Ryan, who told me to say yes to this project;

Jeff, Robin and Ryan, for their help with the reference photos for the original illustrations in this book;

Justin, Pete, Steve, and Tim, my first training partners;

Casey, Frank, Sam, Serge, Stephen, Tim, Tyler, and Will, without whose explorations into Phoenix Earth I would never have found historical European martial arts (HEMA);

Mom, Dad, and Kim, for a lifetime of love and support;

Lisa, for your unflagging belief in me, your support of Swordwind, and your love and friendship; and

Alanna, who made me realize that if I ever wanted to teach my daughter swordplay, I had better start learning. May the sword find its way to your hand one day.

A NOTE ON TRANSLATIONS

I am deeply grateful to the transcribers and translators of historical European fencing treatises whose work has helped make this book—and, indeed, the entire project of historical European martial arts—possible. More often than not, these translations are labors of love produced by nonprofessional translators and offered to the historical fencing community for little or no financial compensation.

One of the most significant contributions to the spread of the modern HEMA movement has been the Wiktenauer (www.wiktenauer.com), a non-profit website dedicated to promoting and disseminating scans, transcriptions, and translations of historical European fencing treatises, free of charge. Several of the translations in this book are taken (with the permission of the various translators) from the Wiktenauer, while others are taken from print editions. Quotations from works in print are made to the page number of the modern print edition rather than the original manuscripts. The format of the Wiktenauer makes it difficult or impossible to cite where in a given author's work a quotation comes from, particularly as many translations draw from multiple manuscript copies of the same text. As a result, where I quote a translation from the Wiktenauer, I have made no attempt at specifying page or folio numbers. Curious readers will find it easy to do a text search on the website. Where I quote a translation from a print edition, I cite the page number of the modern print edition.

In deference to the difficulty of the choices that any translator of historical fencing treatises must make, I have kept untranslated any words that a translator whose work appears in this book has chosen to leave in the original language. As a result, the reader will occasionally be confronted with technical fencing jargon in German, Italian, or Spanish. Editorial or explanatory comments in brackets are also the translator's, rather than my own.

I have tried to select quotations from treatises that minimize the necessity for the reader to understand technical terms, and to clarify these terms when and as they arise. However, there is one set of terms that recurs so frequently that we should discuss it here. Both escrima comun and Bolognese fencing treatises (for more on these terms, see Chapter 2) frequently refer to cuts by reference to whether the cut is forehand or backhand. In escrima comun treatises, a forehand cut is referred to as a *tajo*, while Bolognese treatises use the term *mandritto*. A backhand cut is referred to as a reves in escrima comun, while in Bolognese it is called a *riverso*. Note that these terms hold no matter which hand is holding the sword: a cut that travels from right to left made with a sword in the right hand (i.e., a forehand cut) is referred to as a *tajo* or *mandritto*, as is a cut that travels from left to right made with a sword held in the *left* hand. The meaning of *reves* and *riverso* is likewise relative to the hand holding the sword rather than the absolute trajectory of the sword through space. For two-handed weapons, cuts are denominated *tajo, mandritto, reves,* and *riverso* based on the fencer's dominant hand (e.g., a right-handed fencer who cuts from right to left would be said to be making a *tajo* or *mandritto* even though the cut is forehand with the right hand and backhand with the left hand).

Lastly, the practice of historical European martial arts is constantly evolving, as is our understanding of the treatises from which we work. Any errors in interpreting the words of historical fencing masters that may appear in this book are my own, and do not reflect on the work of the translators or any other person.

It is a Herculean task to translate a technical work into another language, especially when that work is attempting to describe an inherently physical topic (such as swordsmanship) that is best understood with sword in hand rather than with words on a page. Those who undertake it, and particularly those who graciously gave permission for their work to appear in this book, have my profound thanks. Quotations are used with express permission, and may not be reproduced or transmitted without the permission of the translator or relevant copyright holder.

PREFACE

I have had the great pleasure of working with Eric on a number of occasions and was thrilled to learn that he had taken on this project and is sharing his knowledge of medieval and Renaissance martial arts with you. As a student, Eric is open and enthusiastic, endeavoring to rise to any challenge put before him. As a coach, Eric displays his caring nature as he supports his students and assists them in their own pursuits. And now, as a writer, Eric brings his joy and love of the Art to the page, sharing them with you.

It is no small task to take on this kind of project, for each discipline he approaches is distinct and highly detailed, and yet, I think that you will find that he distills these broad topics into the essence of each weapon's use. As each chapter unfolds, he discusses the specifications of the weapon itself, defining the length, weight, and makeup of it as an object. He then discusses the context of the weapon's use, drawing connections between the techniques and advice of a variety of medieval authors. Each chapter concludes with the general impression of the armament just discussed, while making connections to the successive weapon.

While I appreciate Eric's diligent approach to this subject in the main text, I find that my favorite feature of this book are the sidebars he has included to approach the kinds of questions that fans of medieval weaponry often have, but are sometimes reluctant to voice out loud. "Which would win ...?" questions are often the most fun, but frustrating, questions that newer students bring to class. Rather than ignoring or disdaining these topics, he dives into them, allowing us as readers to explore them in a respectful and knowledgeable way.

If you are new to these Arts, I hope that you are able to take your time as you read this volume, and to reflect upon the links that Eric has made between the various medieval masters, but I will forgive you if you find that he has made the topic so approachable that you devour it in one sitting. For those of you who have experience with these Arts, I am confident that you will find his approach to be novel in its whole-hearted clarity, avoiding the trap of unnecessary complexity. And for those of us who teach these Arts, Eric has given us a new way to talk about the Art we love, and we now have a wonderful resource to recommend to our newer students, especially if they have questions about weapons or masters we don't personally study.

—Jessica Finley, author of *Medieval Wrestling: Modern Practice of a Fifteenth Century Art* and Instructor, Ritterkunst Turnhalle

CHAPTER ONE

We're all sword geeks here

> *As a young man I desired to learn armed fighting, including the art of fighting in the lists with spear, poleaxe, sword, dagger and unarmed grappling, on foot and on horseback, armored and unarmored.*
> —Fiore dei Liberi (Fiore, 2017, p. 1r)

This is a book about history, and specifically the history of weapons, but it is likely different from other books you may have read on the subject. Books about weapons tend, in my experience, to fall into two categories. The first are the coffee table books, which present weapons as art. They are filled with oddities such as axe pistols; their pages drip with jeweled hilts and engraved blades. The second are the history books, which present weapons as a means to some other end: how certain weapons and armor influenced the military history of a given period, perhaps, or the sociopolitical consequences of different methods of waging war. These are perfectly valid approaches to the topic of historical weaponry, but neither approach quite sees weapons *as* weapons. Amidst the appreciation of intricate craftsmanship and larger historical significance, a much more basic question remains unanswered: how do you actually use the darn things?

2   THE USE OF MEDIEVAL WEAPONRY

This is a book about that question: how were medieval weapons used? We will take the perspective not of the artist, nor of the general, but of the person actually holding the sword. You've never read a history book quite like this one.

This book discusses the use of eight weapons or weapon combinations from the Late Middle Ages and early Renaissance, selected with an eye toward their iconic status in popular culture. We will cover the use of the longsword; the sword and buckler; the sword and shield; the sword and cloak, dagger, or second sword; knives and daggers; greatswords; spears and other thrusting polearms; and poleaxes, halberds, and other cutting polearms. We will also talk specifically about the use of these weapons in and against armor and in war, and how those contexts do and do not change the way weapons are used.

We will *not* cover mounted combat or combat with ranged weaponry, except tangentially. This is not to marginalize either topic, and I quite understand that the interactions among hand weapons and horses and missile weapons are of significant historical interest. Alas, mounted and ranged combat are simply too far outside my own expertise to do them justice, so their treatment must await another author.

Why does any of this matter? In one sense, it doesn't. You don't have to know how to swing a sword to appreciate it as art. The differences between various nations' martial arts styles have rarely, if ever, changed the outcome of a war. If all you care about are high-level trends and concepts, there is a colorable argument to be made that the ins and outs of weapons usage have no real historical significance.

Of course, medieval martial arts *were* an important part of medieval culture to medieval people. How they used their weapons sheds light on their attitudes, behaviors, and even the way they chose to organize themselves. But to be perfectly honest with you, high-minded intellectual curiosity is not why this book exists.

This book exists because swords are cool.

There. I said it.

The real genesis of this project is in games. When I was a teenager, my friends and I played a lot of tabletop roleplaying games such as *Dungeons and Dragons*. We were storytellers, and we crafted long-form collaborative storytelling experiences that spanned years of real-time, full of intrigue, emotions, and derring-do.

The one part of these games that never quite satisfied me was their combat rules. I didn't just want to know that Kalaraen hit an orc with her longsword. I wanted to know how she actually swung it, and why, and what the orc was doing at that moment. I wanted to know why and in what circumstances a person would even *use* a longsword, instead of some other weapon—and I wasn't satisfied with purely in-game answers.

By my sophomore year of college, I was ready to craft my own rules that would address these perceived shortcomings. There was only one problem: I didn't actually know how to use a sword, let alone how the various types of historical swords really differed from each other from a usage standpoint. A quarter's worth of foil fencing on campus did not bring me any closer to understanding. There were a couple of reenactment-type groups and live-action roleplaying groups nearby that engaged in mock combat, but so far as I could tell, they were just making things up based on their own rough-and-ready experiments. That wasn't satisfying to me; I wanted something more historically grounded. Perhaps, I thought, I could get at the answers I wanted by cobbling together some combination of military history, heroic literature, and weapons archaeology.

So I hopped onto the baby internet and dove into the forgotten part of my university's library, looking for answers. If you're reading these words, I imagine the odds are pretty good that at some point in your life, you've done something similar.

My research led me to Giacomo di Grassi's *His True Art of Defense*, a 1570 work by an Italian fencing master (translated into English in 1594) with step-by-step instructions on how and why to use a variety of weapons. This was a revelation to me. Historical people wrote down how they used weapons? We didn't have to guess?

It turns out, we don't. Or at least, we don't have to guess so long as we can understand centuries-old martial arts books … which actually requires a lot of study, sweat, and bruises. This book is based on those primary sources and the experience of myself and others like me who practice the martial arts they describe.

Nothing you read here is going to turn you into a medieval martial artist. No book can. But you *will* come away with an understanding of what medieval martial arts looked like, a sense of how the experience of using one type of weapon is different from the experience of using another, and the pros and cons of each. This is the book I wished was available when I was in high school. To the spiritual kin of my teenage self: I hope what you read here both satisfies and inspires your curiosity.

And that, dear reader, is the real reason that this book exists. Although this is a book about history, it is not written for the historian. I wanted to understand swords from the medieval user's perspective so I could tell more compelling stories with my friends. If you have ever had similar feelings—if you have ever wanted your fantasy to feel a bit more real, if you have ever wondered what your favorite movie sword fight *ought* to have looked like, if you have ever read the placard next to a sword in a museum and wanted to understand the object behind the glass more deeply—then you are in the right place.

We're all sword geeks here.

CHAPTER TWO

How do we know any of this?

> *And Master Sigmund ein Ringeck, fencing master to the highborn prince and noble Lord Albrecht, Pfalzgraf of Rhein and Herzog of Bavaria had these same veiled and misleading words glossed and interpreted as lay written here in this little book, so that any fencer that can otherwise fight can fully absorb and understand it.*
> —Sigmund Ringeck (Ringeck, 2019)

Before we get into the use of weaponry itself, it's worth asking how we know any of this.

Recall the moment of revelation I experienced when I discovered di Grassi. As it turns out, European fencing masters from at least the 14th century CE actually wrote down how to fence with the weapons of their day. These treatises (I'm going to refrain from calling them "manuals," as many of them don't quite fit the description of true "how-to" books) are not works of history, like a medieval chronicle, nor works of art, like an epic poem. These are technical works about how to fight. A brief example from the Bolognese fencing master Achille Marozzo will suffice to make the point:

> Make your student set himself with his right leg forward, with his sword and his large buckler or targa well extended straight toward his enemy, and with them close together. Make him hold his right hand outside his right knee, with the wrist of his sword hand turned downward toward the ground … This is called "coda lunga e stretta," and it is equally good for attacking as it is for parrying.
>
> Achille Marozzo (Marozzo, 2018, p. 165)

Marozzo's words are clear and to the point. He introduces the reader to technical fencing jargon. From this single excerpt, you already have a sense of how to stand in the guard he calls coda lunga e stretta—and the curious reader of Marozzo's treatise can always refer to his included illustration for further clarification.

Sidebar: Fencing or sword fighting?

Throughout this book, I'm going to refer to people who used weapons as "fencers," and the activity of using weapons as "fencing." Depending on your background, this may seem strange to you. To some people, the word "fencing" connotes skinny athletes dressed in white bouncing back and forth and hitting each other with "swords" that seem more like car antennas. You might be used to differentiating between this activity and "sword fighting."

The English word "fencing" originally referred to the activity of armed self-defense (or defence in the British spelling; hence the c in fencing). I prefer "fencing" to "sword fighting" for two reasons. The first is that it connotes a certain elegance and technicality—more *The Princess Bride*, less *Conan the Barbarian*—and the traditional martial arts of Europe were highly technical. The second is that medieval fencing was oftentimes literally *not* "sword fighting," in that the weapons employed were not always swords. "Fencing" in the historical sense of the word covers the use of any hand weapon at all: swords, yes, but also quarterstaffs, spears, axes, flails, maces, warhammers, sickles, and scythes (yes, scythes). As this book is concerned with more weapons than just swords, "sword fighting" just doesn't seem like the right term.

That said, understanding medieval fencing is not as simple as picking up a book by a medieval fencing master and reading it cover to cover. In part, this is due to the fact that medieval academic thought was very different from modern academic thought, which makes it difficult for modern readers to really follow medieval technical writing. In part, it is due to the fact that medieval books had to be produced by hand, which tended to limit their word count and thus cause them to treat some topics with less detail than we modern readers would prefer (Gutenberg produced his printing press in the 1430s, but books on fencing in the Late Middle Ages were still overwhelmingly handmade).

And in part, of course, it is due to the nature of fencing itself. Martial arts are an inherently physical activity. There are numerous nuances to executing even the simplest action that are most efficiently demonstrated in person. The authors of these works, as fencing masters themselves, would have expected to be able to instruct their students in person, with swords in hand. A correction of form that might take hundreds of words to convey could be accomplished in person simply by moving the student's hand to the correct position. As the anonymous fencing master who contributed to a text we call the Nuremburg Hausbuch wrote:

> [O]ne cannot speak or explain or write about fencing quite as simply and clearly as one can easily indicate and inform it with by hand.
> (Nuremberg Hausbuch)

Despite these difficulties, a serious effort to reconstruct medieval fencing has been underway across the world for the past several decades. This movement is commonly referred to as historical European martial arts, and it forms the foundation of this book. HEMA (for our purposes, anyway; I won't claim to speak for everybody who claims to do HEMA) is (i) the study of surviving European fencing treatises and (ii) the physical reconstruction of the martial arts in those treatises by modern practitioners.

Both aspects are important. As the anonymous author in the Nuremburg Hausbuch said, you can't really learn how to fence without actually fencing. Understanding the treatises requires trying to do what they describe, in the way they describe it. This is the only way to confirm that we understand as much as we think we do from reading. At the same time, what the texts describe remains the touchstone of HEMA. The study of period texts gives HEMA more historical grounding than simply picking up replica weapons and "seeing what works."

Let's say a little bit more about the virtues and limitations of this approach. Experimental archaeology with replica weapons (the "pick it up and see" approach) can let us say, "It seems plausible that weapons could have been used this way." If modern people can do something with accurate replica weapons, it stands to reason that medieval people could probably do it with the real things. Human bodies with a given weapon can only move in so many ways, after all (you'll hear this refrain frequently if you spend any time in or around historical sword-related communities). However, the experimental archaeology approach is limited to making claims about what *might* have been done. What *was* done is inherently beyond the realm of experimental archaeology.

The HEMA approach lets us go beyond the plausible. Treatises allow us to take the critical step beyond "This *might* have been how people fought" to "This is at least one method of fighting that was actually recommended." That narrows our historical uncertainty considerably. It gives us a level of insight that simply isn't available for weapons or periods that aren't covered by extant fencing treatises.

More than this, treatises let us access not just what medieval fencers did with their bodies and weapons, but how they thought about the fight (or, at least, how their instructors wanted them to think about the fight). A practical martial arts system is not about what the human body *can* do. It is more often about what the human body should *not* do (I *can* stand on my head and clench my sword in my teeth, but should I?). It is about *when* the human body should do something ... and, most importantly, it is about *why*. Treatises give us windows into these aspects of medieval fighting psychology.

This is not to say, of course, that we can make the leap from "I saw this in a medieval book" to "all medieval people fought like this." That would be absurd. In fact, the skeptical among you may be saying to yourselves right now, "Wait a minute. *Anybody* can write a book. Why should we believe *anything* these so-called 'fencing masters' say? What if they were frauds?"

That's a good question! My answer is that historical inquiry is not about achieving certainty. It is about decreasing uncertainty. Here are some factors that might make us more confident that a given fencing master knew what he was talking about:

1. **He wrote a book.** This may not count for much, but it's better than nothing. Writing books is neither an easy nor a casual endeavor,

especially in the Middle Ages. People who put in the time and effort to write about fencing may not be much more likely to be an authority than the average man—but they are *somewhat* more likely.
2. **He had a documented history of fighting success.** People who actually win fights are more likely to be authorities on how to win fights than people who lose or never fight at all. It is, of course, entirely possible to win a fight through factors other than skill (not the least of which is sheer dumb luck!). Nevertheless, all other things being equal, I'd rather listen to someone who has won fights with his own two hands than to someone who hasn't.
3. **He was quoted by later fencing masters.** If a master's treatise was significant enough to be worth talking about in subsequent generations, its teachings were probably considered fairly authoritative. This holds true whether a work is quoted favorably or unfavorably. You don't bother to trash talk somebody to whom nobody listens in the first place, after all.
4. **His treatise was republished or copied multiple times.** Multiple publishing runs or a treatise being copied many times (medieval treatises were generally produced by hand and thus had to be copied by hand) indicate that period fencers thought a particular work was worthwhile. This is especially true if the work was republished or copied beyond the author's lifetime.
5. **He belongs to, or founded, a martial arts tradition that spans generations.** For a master to be talked about or read after his own lifetime indicates that he possessed a fair amount of influence and expertise. For a master to be *followed* after his own lifetime suggests that he—or, at least, the tradition about which he wrote—was considered well worth listening to.

None of this eliminates the possibility that the documented martial arts traditions of Europe were an elaborate fraud perpetrated upon history by an international network of charlatans. But it does decrease the likelihood.

Medieval Europe seems to have been rife with different martial arts traditions, not all of which are represented by texts that survived to the present day. Some of them may never have been recorded in written form at all. As we've discussed, writing isn't necessarily the best or most obvious way to transmit physical knowledge. Others may have produced treatises that simply did not survive the centuries—perhaps because nobody thought the tradition's teachings were good enough to preserve, but more likely due to the vagaries of history. There are a lot of ways for a book to get lost in hundreds of years.

Whether a tradition didn't produce treatises in the first place or those treatises simply didn't survive, we modern people are limited to those treatises that did survive. As we'll be referring to these a fair amount in the chapters to come, let's discuss them a bit now. There are five principal traditions and authors from which we're going to draw:

- **Kunst des Fechtens:** Kunst des Fechtens (KDF) is the fencing system of Johannes Liechtenauer, a famous fencing master of the Holy Roman Empire. Liechtenauer lived in the late 14th or early 15th century, and left behind his teachings in a cryptic poem called the *zettel* ("epitome" or "recital"). KDF grew from an apparently secret society of swordsmen in the 15th century to an important part of the curriculum of the Marxbrüder, the only fencing guild in the Empire with the official endorsement of the emperor. If you've heard of "German longsword," the odds are good that the speaker was referring to KDF. Masters in this tradition include Johannes Liechtenauer, Sigmund Ringeck, Peter von Danzig, Hans Seydenfaden, Martin Huntzfelt, Ott Jud, Jud Lew, Andre Lignitzer, Paulus Kal, Joachim Meyer, the anonymous author of a fencing treatise in the Nuremburg Hausbuch, and (possibly) Hans Talhoffer and Johannes Lecküchner. Weapons treated include the longsword, spear, dagger, langes messer, sword and buckler, poleaxe, and longshield.
- **Bolognese:** The Bolognese tradition is centered on Bologna, Italy, which housed a documented fencing academy since at least the early 15th century. Unlike KDF, Bolognese does not have a single progenitor or high master from which it claims descent. It was nevertheless a very popular style. Bolognese fencers included some of the great mercenary captains of Italy and the fencing master to King Henry IV of France. Masters in this tradition include Filippo Dardi, Guido Antonio di Luca, Antonio Manciolino, Achille Marozzo, and Giovanni dall'Agocchie. Weapons treated include the sword alone, sword and buckler, sword and shield, sword and dagger, sword and

cloak, two swords, cloak and dagger, dagger alone, greatsword, and a variety of polearms.
- **Escrima Comun:** Escrima comun is the name given to fencing from Spain and Portugal prior to 1624, when Luis Pacheco de Narvaez became the royal fencing master and initiated a kingdom-wide program of wiping out the old methods of fencing in favor of *la verdadera destreza* (LVD), the style that he himself practiced. Unfortunately, Pacheco's purge—complete with literal book burnings of older fencing texts—seems to have been remarkably effective; only one text in the escrima comun tradition is known to have survived. Fortunately, Pacheco was enough of a zealot to describe the techniques of older Spanish masters in some detail (all in the name of debunking their "inferior" fencing), which allows us to describe escrima comun as a coherent tradition with reasonable confidence. Masters in this tradition include Domingo Luis Godinho, Jaime Pons, Pedro de la Torre, and Francisco Román. Weapons treated include the sword alone, sword and buckler, sword and shield, sword and dagger, sword and cloak, two swords, and greatsword.
- **Armizare:** Not a tradition as such, armizare is the name given to the art of Fiore Furlano de'i Liberi (usually shortened to Fiore dei Liberi), a 14th-century Italian knight. Fiore's students seem not to have carried on his teachings (although the work of Philippo Vadi in the late 15th century bears a striking resemblance), he wrote his treatise after a long life as a soldier of fortune and fencing master to various famous mercenary captains. If you've heard of "Italian longsword," the speaker probably meant armizare. Fiore's work is especially interesting to modern practitioners, as his treatise is both exceptionally clear for a medieval work and self-illustrated. Fiore treats the longsword, marshal's baton, dagger, spear, and poleaxe.
- **Pietro Monte:** Pietro Monte was a famous 15th-century mercenary captain who wrote four separate treatises on military matters. Of particular interest to HEMA practitioners is his *Exercitiorum Atque Artis Militaris Collectanea*, or *Collection of Military Arts and Exercises*. The *Collectanea* contains Monte's thoughts on the use of a wide range of weaponry, as well as details about how to train and exercise that few other fencing texts commit to print.

> **Sidebar: Martial arts by any other name**
>
> European martial arts traditions have generally not come down to us with explicit names. A given master's system is most often referred to within his work simply as "the art of fencing," "my art," or similarly. Indeed, fencing masters sometimes seem at odds with each other, and even themselves, as to whether distinct martial traditions even existed. For all the self-evident differences among European treatises in phrasing, technical vocabulary, and tactics, it is easy to find statements along the lines of "there is but one art of the sword" (from the Nuremberg Hausbuch) or claiming that other fencers are bad simply because they don't understand the true art—singular—of fencing (Fiore is particularly fond of this phrasing, though he is by no means the only one). The names given to martial arts traditions in this book are conventional among modern HEMA practitioners. They should not be taken as strictly historical.

The historically literate among you may have noticed that not all of the masters named above lived, strictly speaking, in the Middle Ages. In fact, they cover quite a range of time: Fiore wrote in 1409, while Godinho wrote in 1599. You may well be asking yourself: if this is a book about medieval weapons, what's with all the Renaissance authors?

The periodization of history—particularly the divide between the "Middle Ages" and the "Renaissance"—is a modern convention. Medieval martial arts traditions did not disappear from Europe simply because later scholars would declare that Europe had entered a new period of history. The writings of "Renaissance-era" masters in these "medieval" traditions are thus relevant to the study of medieval fighting. Indeed, there are some arts with roots firmly in the Middle Ages for which there are no known treatises until the Renaissance. Filippo Dardi began teaching fencing in Bologna in 1412, but the earliest surviving Bolognese treatise is that of Antonio Manciolino, published circa 1523.

Moreover, arts such as Bolognese and escrima comun were considered markedly different and "old-fashioned" in the Renaissance compared to contemporaneous "modern" arts. Consider the case of Angelo Viggiani, who wrote in the 1550s. Viggiani's treatise is in the form of a classical dialog between one Rodomonte, who represents Viggiani's "new" style of fencing, and the Conte d'Agomonte, a knight who seems to have been trained in the Bolognese style. In the following excerpt, the Conte demonstrates a classic Bolognese guard to Rodomonte, who lambasts it as absurd:

> RODOMONTE: Open well your eyes, and watch what I do: place yourself, Conte, in whatever guard you wish.
> CONTE: Look, I place myself in cinghiara porta di ferro [the wild boar's iron door].
> ROD: Oh, by your faith, Conte, don't give me these bizarre names of guards of yours, please stop saying these code lunghe distese [extended long tails], your falconi [guards of the falcon], porte de ferro larghe [wide iron doors] or strette [narrow iron doors], and such strange fantasies!
>
> Angello Viggiani (Viggiani, 2018, p. 25)

Viggiani goes on to provide a much more "scientific" (by his estimation, anyway) system of guards, with far less fantastical names. You may be familiar with the modern fencing convention of numbered guards (prime, seconde, tierce, quarte, etc.). This practice began in the Renaissance—but the old ways persisted, and indeed coexisted with newer fencing systems, for quite a while.

What about earlier works? The earliest master we've considered so far is Fiore (Liechtenauer may have been Fiore's contemporary, but was almost certainly no older), and he was born in the 14th century. What about the rest of the Middle Ages? There's nearly a thousand years of medieval history by the time the 14th century rolls around. What about all of that? What about Vikings and Normans and Charlemagne? What about Richard the Lionheart and William Marshall, Jean II Le Maingre and Geoffroi de Charny?

The sad fact of the matter is that fencing treatises simply cannot carry us that far back in history. The earliest known European fencing treatise is the so-called Walpurgis Fechtbuch (sometimes known as I.33, its former catalog number at the Royal Armouries in Leeds in the United Kingdom; as I.33 is no longer the text's current catalog number, I shall refrain from this usage), which dates only to the 1320s. Undoubtedly, European martial arts stretch further back than that. Somebody taught Fiore, after all! In fact, he names a few of his teachers; sadly, none of them are identifiable historical persons. Nor do other medieval arts speak as if they have invented martial arts out of whole cloth in the 14th or 15th centuries. The Nuremburg Hausbuch notes that

> [T]here is but one art of the sword and it may have been *invented and thought out many hundred years ago*. And this is the basis and core of all of the arts of fencing and Master Liechtenauer had internalized and applied this quite completely and correctly (emphasis added).
> (Nuremberg Hausbuch)

In other words, the martial arts we have documented were not necessarily new. According to the Nuremburg Hausbuch, even Liechtenauer, allegedly a great martial arts synthesizer and systematizer, merely *internalized and applied* a preexisting art. The sources we draw on date from the 15th and 16th centuries, but they should be understood as describing an older practice.

How *much* older? That's a much trickier question. It seems very unlikely that armizare looked too radically different from other Italian arts in 1409 (the date of one copy of Fiore's treatise, which has illustrations drawn by the master himself). Traditions tend to diverge over time, after all, rather than bursting onto the scene as something completely new. If that's true, it seems likewise unlikely that armizare would have looked very radical even in 1399. Perhaps even in 1390 it would seem mostly familiar. But what about 1300? 1200? 1100? Would armizare, Bolognese, or KDF look familiar to William the Conqueror? Would *any* 15th-century fencing?

The question is not wholly unanswerable. We can compare fencing from our 15th-century treatises to artwork from earlier centuries, for instance, and see what guards or positions seem to be consistent across time. We can observe changes or continuity in weapon and armor design, and speculate as to how that might drive changes in martial practice. There are historical fencers whose passion is this very project. However, it is the 15th and 16th centuries, and the documented traditions of those centuries, that anchor our understanding of medieval martial arts. Because of that, it is to those centuries and traditions that we shall confine ourselves in this book.

Although we will be drawing from all these systems and a few others, I do not wish to be understood as claiming that KDF, Bolognese, escrima comun, and armizare are all interchangeable. They are each distinct historical artifacts, worthy of study in their own right (and for those of you who are already dreaming of taking the best parts of each of them and creating some sort of uber-system ... study one for 20 years or so, and you might have a reasonable claim to understanding what the "best" parts are of that *one* system).

However, as we have already discussed, the core of a martial arts system is how, when, and why. It would take an entire book to discuss the how, when, and why of even a single medieval fencing system, and that book would be ... compendious, to say the least.

Our approach is going to be different. We will certainly touch on the how, when, and why, but our main concern in this book is even more basic: *what* did people do with medieval weaponry? That is an answer that has a lot more commonality across systems. This technique-centric survey approach will not serve to teach you how to fight, and I do not recommend it for actual fencing lessons. It is ideal, however, for getting an understanding of the dilemmas a fencer must solve, and the tools that experienced martial artists across Europe used to solve those dilemmas. With the major sources and general approach now introduced, let us turn to the weapons themselves.

CHAPTER THREE

Of royal blood: longsword

> *I'm the sword, deadly against all weapons. Neither spear nor poleaxe, nor dagger can prevail against me. I can strike long or short, or I can be held in the half sword grip to move to the Close Range Game. I can be used to take away the opponent's sword, or move to grapple. My skill lies in breaking and binding. I'm also skilled in covering and striking, with which I seek always to finish the fight. I'll crush anyone who opposes me. I'm of royal blood.*
>
> —Fiore dei Liberi (Fiore, 2017, p. 25r)

The longsword occupies a peculiarly privileged position in modern HEMA, and especially in modern medieval HEMA. We shall therefore begin our discussion of medieval weapons with the sword that Fiore describes as "of royal blood."

First, to define our terms: I am using "longsword" as a modern term, invented by modern historical fencers for our own convenience. It is not a historical term. There *is* no single historical term for the weapon we are discussing, which is the reason we HEMA practitioners had to invent our own. What we really mean by "longsword" is a sword such as might be used by a practitioner of armizare or KDF—the late medieval systems from which most of this chapter will be drawn. To be used in the manner that both armizare and KDF describe, such a weapon must have a grip long enough to accommodate two hands, be light enough to wield in one hand, and be small enough to be able to draw from the belt (albeit sometimes with special suspensions that give the scabbard more room to move backward as the sword is drawn). In practice, such a sword tends to have a blade of 33" to 40" long, with an additional 7" to 10" of hilt, and weighs around 3–4 pounds. You may be familiar with the weapon under the names "hand-and-a-half sword" or "bastard sword" (a term that appears in medieval armory lists, but is difficult to define with certainty).

Sidebar: Bring me my long sword!

The choice of words we use in modern English to refer to historical weaponry is fraught with difficulties. For weapons in current use, the words we use to refer to them can make many assumptions. If I say the word "pistol" to you, you probably picture a modern self-loading handgun; you can safely assume that I am not speaking of a 16th-century wheellock. If I say the word "sword" to you, you can make no analogous assumption. For this reason—and because modern people have a positively Victorian obsession with categorizing the past—we need to make choices about how to refer to swords that the people who actually used them would not have had to make. For this reason, it's always important to be aware that the modern names we use for swords *are* modern names.

The longsword is a perfect example. Modern historical fencers refer to longswords as longswords because they are the sort of weapon used in German fencing texts describing the "art of the long sword." "Long" in Middle High German has the connotation of being far away, not just literal length (compare "go long" in English sports terminology). The "art of the long sword," to German-speaking medieval fencers, was the art of using the sword at long range, not the art of using a sword that is longer than other swords. German fencers did not generally refer to their longswords as long swords. Most of the time, they simply referred to them as "swords," and when they do seem to be using more technical terms, they most often characterize swords by intended use (for instance, Andre Paurñfeyndt, a 16th-century KDF master, says that his art of the long sword can be used with the "battle sword" and the "riding sword," "among many others") rather than by physical description.

> Likewise, when Shakespeare has old Capulet cry, "Bring me my long sword!" in *Romeo and Juliet* Act I, Scene I, he is not calling for a "longsword" in the sense we are using that word in this book. He is calling for his *big* sword (probably a rapier, given the setting—possibly one of several that he owns). "Bring me the big one!" is closer to his meaning. When the *Dungeons and Dragons* of my youth labeled knightly one-handed swords "long swords," it was doing so to distinguish them from "short swords"—in fact, by the standards of the game, a "long sword" was not even a particularly long sword.
>
> There's nothing wrong with any of these choices. We simply need to be aware that sword terminology is fraught with complexity, and that the words we think of as "technically correct" may not be. Oftentimes, there is more than one—or no—"technically correct" way to refer to swords.

As I mentioned earlier in this chapter, the longsword is (by definition) the archetypal sword used in both armizare and KDF, each of which devotes a substantial amount of its corpus to the weapon. This wealth of available material has, in turn, made it extremely popular among HEMA practitioners; if a HEMA school is involved with the Middle Ages at all, it is a safe bet that it teaches the use of the longsword. Despite this, there is little evidence to suggest that the longsword was especially popular in the Late Middle Ages themselves. It is most often depicted in art wielded by high-status (and therefore rare) warriors. Whence, then, the apparent fascination of period martial artists with the longsword?

Part of the answer, undoubtedly, was a fascination with the weapons of the elite. But there are pedagogical reasons to teach the use of the longsword as well.

In KDF, the longsword almost seems to be a sort of beginner's weapon. Liechtenauer discusses the use of the sword before he discusses the use of any other weapons. Fiore introduces the reader to grappling, the marshal's baton, and the dagger before the longsword, but he also makes a point of the longsword's versatility:

> I can strike long or short, or I can be held in the half sword grip
> to move to the Close Range Game. I can be used to take away the

opponent's sword, or move to grapple. My skill lies in breaking and binding. I'm also skilled in covering and striking, with which I seek always to finish the fight.

(Fiore, 2017, p. 25r)

This versatility may explain why KDF and armizare seem so enamored of the longsword as opposed to other types of sword. It occupies a useful middle ground between large and small weapons. For KDF, it provides a good foundation for teaching the principles of armed combat that can then be adapted to smaller one-handed weapons and to larger polearms. For armizare, it provides a bridge as the student learns to apply the principles learned with small weapons—grappling, dagger, and the marshal's baton—to larger weapons, without jumping immediately from these small weapons to large polearms.

Speaking of size, there's something we should clear up about longswords: they are not, in actual use, particularly long swords. Longswords do typically have blades that are a few inches longer than contemporary one-handed swords (which we will discuss more in Chapters 4 and 5). This length, however, does not necessarily translate into superior reach. Because the longsword is held in two hands (in most cases—we will discuss exceptions later in this chapter), it is connected to both shoulders. Advancing the dominant shoulder necessarily pulls the non-dominant shoulder away from the opponent. When striking with a two-handed sword, there is a limit to how much the fencer can do this before the non-dominant hand is pulled off the weapon's handle entirely. Thus, a longsword fencer's shoulders need to stay relatively square to the opponent (making some allowance for the rotation of the hips and shoulders as a part of proper striking mechanics). By contrast, a one-handed sword allows a fencer to advance the sword-side shoulder slightly further toward the opponent (you can still see this, taken to its logical extreme, with the side-on stance of modern fencers). As a result, a longsword of a given length actually has a few inches *less* reach, practically speaking, than a one-handed sword of equivalent length. While no treatises say as much, it seems plausible that this is among the reasons that longsword blades tend to be a few inches longer than otherwise equivalent contemporary one-handed swords.

If the two-handed grip shortens a fencer's effective reach, we might well ask what it's good for. When I was younger, I imagined that the main purpose of having two hands on a sword was to create more powerful swings. As it turns out, this is not especially true. Obtaining scientifically reliable numbers for this sort of thing is fraught with difficulty, but in my personal experience, longswords only hit perhaps 20% harder than equivalent one-handed swords (most of the power of a cut comes from the core muscles regardless of sword type). What the two-handed grip does provide, however, is superior leverage.

By "leverage," I am referring to a longsword's ability to push another weapon out of the way. Imagine this scenario: two fencers throw simultaneous cuts at each other from their right shoulders. The swords will clash between the fencers with their tips pointed over their heads. Leverage is the quality that will permit one sword to bring its point down to stab the opponent in the face ("Stab him in the face" is a phrase that recurs throughout the KDF corpus, a delightfully prosaic chorus for what is otherwise in many ways a highly esoteric art).

There are many factors that dictate the leverage one sword has on another, including where the swords are crossed (a fencer can exert more leverage closer to the hand than close to the tip), how the edges are oriented to one another (a fencer can exert more leverage with the edge than with the flat), the fencer's body position, and more. All things being equal, however, a sword held in two hands will have more leverage than a sword held in one. This gives the longsword an inherent advantage against one-handed swords whenever their blades are touching.

This is not to say that a one-handed sword cannot parry or control a longsword. Any sword is *capable* of parrying a longsword, or even a greatsword. However, the greater the leverage differential between weapons, the closer to the hand the weaker weapon needs to cross blades. This places the sword hand and forearm in greater danger and tends to make it more difficult to quickly bring the weaker weapon to bear in a counterattack. Crossing the opponent's blade close to the hand is generally good form in fencing anyway, but a longsword's superior leverage essentially allows it to "cheat" this rule in a way that is frequently advantageous.

This leverage is far more important to a longsword than any power lent to its cuts by its two-handed grip. This brings us to another point about longsword fencing: it features a surprisingly large proportion of thrusts. Here, it may be instructive to compare the first longsword techniques in both KDF and armizare. The first technique (or "play," as it is often called in HEMA) that KDF teaches for the longsword is recorded by Liechtenauer's zettel in this couplet, as glossed by the anonymous fencing master sometimes known as Pseudo-Peter von Danzig:

> *Who strikes at you above, the Wrath Stroke threatens him with the point.*
> Gloss—*Note:* the *Zornhau* breaks with the point all *Oberhau* and yet is nothing but a simple peasant stroke. And do it thus:
> When you come to him in the *Zufechten* and he strikes from his right side above to your head, then strike also from your right side from above without any parrying wrathfully onto his sword at the same time as him. Then if he is soft at the sword, shoot the point straight in forward and long, and thrust to his face or breast. Thus plant upon him.
> Pseudo-Peter von Danzig (Von Danzig, 2010, p. 113)

In this play, the attacking fencer throws an overhead descending cut at the opponent. The opponent defends not by parrying but throwing an overhead descending cut in return, at an angle (and, as practical experiment shows, often with a small step to the side) that deflects the incoming cut and leaves the point of the sword in the attacker's face. From here, the defender stabs the attacker in the face. The play goes on to describe follow-up options in case the opponent is sufficiently on the ball to defend against the thrust, but half of those follow-up options still end with the opponent getting stabbed (the remainder end in cuts).

Fiore's first play of the sword in two hands follows a similar structure. He writes:

> This Master who is crossed at the point of his sword with this player says: "When I'm crossed at the points, I quickly switch my sword to the other side and strike him from that side with a falling blow to his head or his arms. Alternatively, I can place a thrust into his face, as the next picture will show."
> (Fiore, 2017, p. 25r)

Here, two fencers have ended with their swords crossed (Fiore doesn't specify how, but one easy way for this scenario to arise is that both throw simultaneous overhead cuts at each other). From this situation, Fiore writes, either fencer can lift the sword and throw a descending cut to the head from the other side of his opponent's sword with a small step ... or quickly tilt the point into the opponent's face.

Both these plays give equal weight to cuts and thrusts. That is the point: not that longsword fencing is all thrusts, as if it were two-handed rapier play (though I can confidently say that the "two-handed rapier" description will occur to anybody who spends any length of time using a longsword according to historical sources), but rather, that authentic longsword fencing features an even mix of cuts and thrusts.

> ### Sidebar: Why did swords have two edges?
>
> The longsword is, like many European swords, double-edged ... but what is the second edge for? This is another aspect of the longsword that is frequently misunderstood. The back edge is usually called the "false" edge in Italian and Spanish texts, and the "short" edge in German texts, regardless of its literal length (the word "short" in German connotes closeness as well as shortness; the "short edge" is literally that edge of the sword that is closer to you than the forward, or "long," edge). In this book, I will call it the back edge in an attempt to be system-neutral.

The back edge of a longsword is more than a backup for the front edge. Rather, the back edge provides another vector of attack.

German and Mediterranean schools of swordsmanship use the back edge in distinctly different ways. Italian longsword systems use the back edge mostly in rising cuts: a fencer cuts down with the front edge, for instance, and throws a follow-up cut along the same trajectory using the back edge. Alternatively, an Italian fencer may begin with the point toward the ground and throw a rising cut with the back edge—for instance, to deflect an incoming attack—that sets up the sword for a powerful downward counterattack with the front edge. This use of the back edge is fundamentally about conserving time and motion: it permits cuts to be chained together more quickly than would be possible if the sword had to be turned over between cuts to bring the front edge to bear again. The most common targets for these types of attacks are incoming attacks or the opponent's hands.

Rising back edge cuts do exist in German swordsmanship, but German fencers tended to use the back edge to attack high targets in horizontal or descending trajectories. This use of the back edge does not necessarily take any less time; rather, it opens up attack trajectories that would not otherwise be open or as easy to access.

The zwerchau is a good example. The archetypal zwerchau looks like this: the fencer begins with the sword over the right shoulder. The fencer then turns the sword in the right hand by pushing on the crossguard with the right thumb, so that what had been the back edge is facing to the fencer's right, and the right hand is cradling the weapon. The fencer then cuts horizontally, ending with the sword on the left side, at or above head height, and the point facing forward.

It is, of course, possible to make a horizontal head-high cut with the forward edge. It is difficult to do so ending *over* the head, however, while a zwerchau (performed with the back edge) can be raised quite high over the head. By permitting the fencer to raise the arms above the shoulders while cutting horizontally, the zwerchau permits a fencer to cover the head against descending strikes and attain greater leverage against an opponent's blade by crossing it near the tip, and none of these benefits are possible to realize without the use of the back edge.

Indeed, as a general rule, a failed cut with a longsword can be transformed into a thrust, and vice versa. While this is true of all swords, it is particularly true for longswords, for longswords are exceedingly maneuverable weapons. One reason for this is the two-handed grip, which allows a longsword fencer to "steer" the weapon by pushing with one hand and pulling with another (this is, by the by, a thoroughly suboptimal way to power a longsword *cut* for reasons that are beyond the scope of this book; interested readers are directed to Michael Edelson's *Cutting with the Medieval Sword* for a thorough explanation). In addition to this, longswords are exceptionally light. They do tend to weigh more than equivalent one-handed swords in absolute terms, but not by very much (the difference in size, remember, is usually only a few inches)—and, since the longsword is a two-handed weapon, it is *proportionally* lighter than most one-handed swords.

This light relative weight and the two-handed grip mean that longswords can be maneuvered and redirected with exceptional swiftness. Most medieval swords, while not heavy in absolute terms, are proportionately heavy enough that they benefit from preserving the weapon's angular momentum. That is, it is usually faster and more efficient to redirect a sword with flowing and wheeling circular cuts than to stop its momentum and reverse course. Longswords are among the few swords that can be stopped dead and redirected in the middle of a fencing exchange, which leads to a much more staccato rhythm to the weapon. As a consequence, the longsword can be surprisingly deceptive, snaking its point or edge around an opponent's guard to strike parts of his body that he thought were defended. As Liechtenauer says in his zettel, against a fencer who has learned how to circumvent parries in this manner, "I say to you truthfully: no one can defend himself without danger" (Von Danzig, 2010, p. 98).

This form of maneuverability is an important part of longsword fencing, so an example may be in order. Sigmund Ringeck gives the following example:

> When you cut in from above with the wrath cut or otherwise, if the opponent parries you with strength, then "Indes" shove your sword's pommel under your right arm with the left hand; and against their sword with crossed hands, strike the opponent across their mouth from behind their sword's blade between the sword and the opponent.
>
> (Ringeck, 2019)

Here, the fencer attacks with a descending cut from the right side, and the opponent defends by parrying to the opponent's left. This places the opponent's blade in the opponent's upper left quadrant, between the attacker and the intended target. In defeating the attack, however, it *also* provides a fulcrum that the attacking fencer can use to advantage. As the swords clash, the attacking fencer shoves the left hand (the one nearest to, or even holding, the pommel) to the right, under the right hand, thereby crossing the hands and orienting the sword so that it points over the opponent's right shoulder and now stands between the opponent and the opponent's sword. The technique finishes with a horizontal cut to the opponent's face.

In the above example, the fencer uses the pommel as a steering device, but another benefit of the longsword's maneuverability is increased ability to use the pommel for direct offensive action. Fiore gives an example that can follow from the same crossing of swords as the example we just discussed from Ringeck:

> This is another play that follows from the crossing of my teacher. And from that crossing I can make this play and all of the others that follow. In this play I grip my opponent at the elbow as shown, and then strike him in the face with the pommel of my sword. After that I can also strike him in the head with a falling strike before he has a chance to make cover against me.
>
> (Fiore, 2017, p. 28r)

In Fiore's example, the fencers begin as before, with the attacking fencer striking from his right and the defending fencer parrying. As before, the attacking fencer uses the opponent's sword as a fulcrum. Now, though, the attacking fencer rotates his sword around the defending sword so that his own sword points directly to the rear, forming a shield between the attacker and the opponent's sword, with the pommel facing forward (Fiore adds that at this point the attacker's left hand can be placed on the opponent's wrist for extra control of the opponent's sword). The fencer then smashes the pommel into the opponent's face.

This, however, does not end the technique. Note what Fiore says after the pommel strike: "After that I can also strike him in the head with a falling strike before he has a chance to make cover against me." Is this overkill, or is it reasonable to expect a fencer who has been struck in the face by a steel pommel still to be capable of fighting?

OF ROYAL BLOOD   27

Pommel strikes are eye-catching techniques and tend to grab the imagination. They are also relatively easy to execute in a competitive environment. As a result, they are somewhat overrepresented in HEMA fencing relative to the place they occupy in the treatises themselves. Pommel strikes are worthy of discussion, but it's important not to overstate their impact. Fiore writes of the pommel strike to the face:

> Make this strike quickly if his face is unprotected, and you'll certainly hurt him. I can tell you from experience that with this strike you'll have him spitting out four teeth. From here, if you wish, you can also throw your sword around his neck, as my fellow student will show you next.
>
> (Fiore, 2017, p. 28r)

Importantly, Fiore does not say "from experience" that the pommel strike will knock an opponent out, crush his skull (although I can add from my own experience that it can certainly dent a fencing mask), or otherwise end the fight. Indeed, he explicitly notes that a follow-up may be necessary.

28   THE USE OF MEDIEVAL WEAPONRY

This is often unintuitive to modern people, who tend to have more experience with blunt impacts than with the effects of blades. We can well imagine the bone-crunching impact of a sword pommel to the face, and I certainly don't wish to minimize what it means to be hit so hard as to lose four teeth from a single blow. At the same time, it

is important to remember that the point of a sword can completely penetrate a human torso, and that the edge of a longsword can dismember or decapitate a man in a single blow. Debilitating as a pommel strike may be, it does not compare to the sheer butchery that a sword can effect with its blade.

No wonder, then, that pommel strikes are a distinct minority among longsword techniques. Nevertheless, there are two other families of pommel strikes that we should discuss, if only to dispel some popular myths about them. The first is holding the blade with both hands in order to strike with the pommel like an improvised warhammer. This technique goes by a number of names, perhaps the famous of which are *mortschlag* and *mordhau* (both translating roughly to "murder blow"). This type of attack appears in treatises exclusively with respect to fencing armored opponents, so we'll discuss this technique more when we discuss armored fencing in Chapter 12 (where, as we will see, they *still* form a minority of techniques).

We should also pause to discuss an esoteric technique from the so-called Gladiatoria Group of treatises, a set of 15th-century German manuscripts that discuss formal duels in full armor, equipped with spear, longsword, small shield (buckler), and dagger. The Gladiatoria contains this unusual technique, in which one duelist unscrews the pommel of his sword and throws, then rushes the opponent with sword or spear.

This technique is so strange that it has spawned a minor internet following (assisted, I suspect, by the colorful but misleading translation, "If you wish to end him rightly …"). A few items are worth noting, though. The first, of course, is that the pommel throw itself is not expected to "end" the opponent at all; rather, it's a distraction to permit the duelist to close with sword or spear. In this sense, it is like every other pommel technique: merely an opener to permit a finishing move using a blade. The second is that such a technique would require a specially-made dueling sword. 15th-century pommels were not normally secured with screws at all, and in any case, unscrewing the pommel of a sword at the start of a duel would presumably require a quick-release type of screw similar to a bayonet lug. Judging by the lack of extant antiques with screw-off pommels, however, this technique cannot have been especially widespread (the Gladiatoria Group is not unique in recommending specialized dueling swords; Fiore suggests several as well).

30 THE USE OF MEDIEVAL WEAPONRY

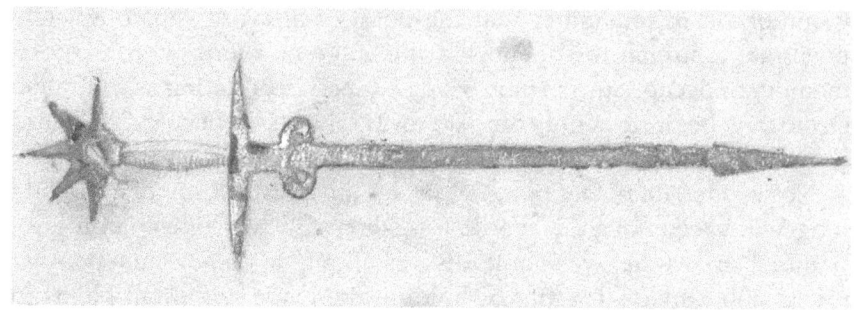

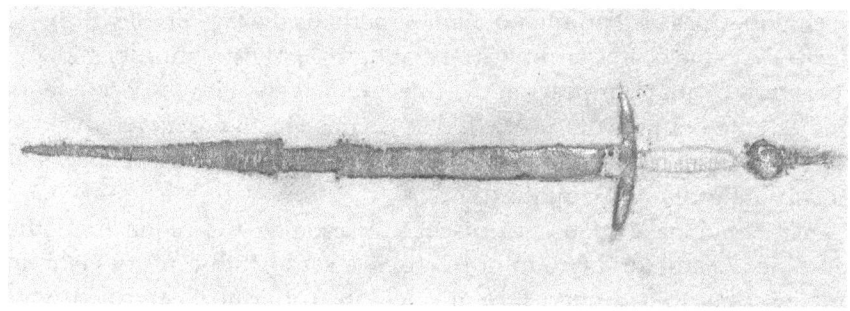

In contrast to pommel strikes, which are prevalent in modern longsword fencing despite being a fairly unimportant family of techniques, grappling is somewhat underrepresented despite being significantly more important. The reason why is not difficult to see: grappling is a highly technical art in which few modern fencers have a preexisting background, while pommel strikes are straightforward and easy to execute. What is true for modern fencers, however, was not necessarily true for medieval fencers. Longsword treatises assume a preexisting familiarity with wrestling that most modern fencers simply do not have, and cannot quickly acquire. As a result, it is common for modern longsword fencers either not to have the skill to execute historical grappling techniques at all, or not to have the skill to execute them *safely*, causing them (hopefully) to refrain from the attempt. A poorly executed grappling technique in a fencing match can easily cause serious and even permanent damage to an opponent's joints, which might be an acceptable result in many historical contexts but is frowned upon in modern competition.

Notwithstanding the limitations of modern practitioners, grappling forms an important minority of historical longsword techniques. The word "grappling" can be confusing, so we should stop to define the term. "Grappling" covers the entire range of techniques in which a fencer lays hands on his opponent or his opponent's weapon. This can be as simple as pushing off an opponent's arm to spin away from an opponent trying to execute a close-range technique, or as flashy as tossing an opponent high over the hip to hit them with the planet, as one of my judoka friends puts it. The historical term includes both holds and throws as well as punches, kicks, and other strikes.

Grappling at the sword is especially important in armored fencing, but even in unarmored fencing, it is fairly common for fencers to end up close enough to each other that a grappling technique is an option. And it is an option, not a requirement: virtually no scenarios require grappling to offend the opponent. If neither opponent wishes to grapple, they can always separate by mutual unspoken agreement. It is also true that in many scenarios, even at close range, it is possible to wound an opponent with the sword. The longsword's length is actually an asset here, as the blade is long enough to comfortably seize with the non-dominant hand, effectively choking up on the weapon (we'll discuss this more in Chapter 12).

Despite the possibility of escaping or declining a potential grapple, we should not downplay the potential lethality of grappling techniques. While pommel strikes are typically presented in treatises as requiring follow-up, grappling techniques are frequently presented as "fight finishers." In part, of course, this is due to the fact that many grappling techniques end with a blow (or repeated blows) from the sword. One of the simplest longsword grapples is to reach inside an opponent's arms with the left hand once swords are crossed, and wrench the opponent's arms out and toward the ground. This gives the fencer temporary control of the opponent's sword arm, and permits the fencer to strike with the longsword one-handed without opposition (the majority of one-handed longsword techniques are grapples of some sort). Although the fencer is technically grappling, the technique finishes with the blade. Naturally, this is considered lethal.

Even throws, however, are usually presented as at least potentially lethal. To take one of many examples from Fiore:

> This play follows on from the previous one, where the student struck his opponent multiple times while using his left arm to keep

the opponent's arms and sword pinned. Now I drive my sword into my opponent's neck as depicted. Then I throw him to the ground to complete the play.

(Fiore, 2017, p. 29r)

Fiore says that the throw "complete[s] the play"; i.e., ends the technique. Pseudo-von Danzig offers insight into a potential reason why, when he describes a throw by saying "grasp him thus on your right hip and throw him before you backwards on his head" (Von Danzig, 2010, p. 127). A throw that results in the opponent falling onto the back of his head could certainly result in incapacitation or even death.

Throws are a remarkably versatile family of techniques. They can result in a sharp blow to the back of the head (as in the example quoted above) or a disabling lateral wrench of the opponent's knee (even in a modern competitive context, where injuring one's opponent is not the goal, grappling techniques can result in joint injuries). At the same time, grappling can also be modified to throw an opponent to the ground with slightly less force or result in a pin.

It is important to remember that not every sword fight in the Middle Ages was to the death. Especially in a "self-defense" context, a fencer might well wish to end an armed confrontation without killing or permanently maiming the opponent, and grappling techniques offered a way to do that. One of my personal favorite longsword grapples neatly shows how grappling can end a fight lethally or non-lethally at the fencer's discretion. This comes from Jud Lew, a KDF master:

> If one takes you by the neck to your right side, then release your sword from your left hand and thrust his sword from your neck with your right, and step with your left foot against his right side before both of his feet, and drive with your left arm over both his arms nearby the hilt, and drive him to dance or stab him below between his legs to the maker.
>
> Jud Lew (Lew 2019)

This grapple is a defensive technique against a slice to the neck. The fencer executing the grapples finds the opponent's blade against the right side of the fencer's neck, with the fencer's own sword outside to the right. The fencer defends against the slice by inverting the fencer's own sword so that the pommel points toward the sky and the blade hangs down, bringing it between the opponent's blade and the fencer's own neck in the process, then turns to the right and wraps the left arm around both of the opponent's arms (this is the same setup as the armizare throw we looked at above; both fencers are facing the same direction). With the opponent's arms thus trapped, the fencer presents the point of the longsword to the opponent's "maker" (i.e., his genitals). From here, Lew says, the fencer can make the opponent dance, like a medieval cowboy shooting at the opponent's feet—or, if the opponent continues to struggle, stab the opponent between the legs. The grapple can end without loss of life, but does not compromise the fencer's ability to escalate to lethal force if necessary.

Before we leave the longsword behind, we should discuss its use against multiple opponents. The historical advice in this regard is fairly clear: don't use a longsword against multiple opponents. Andres Juden (Andres "the Jew"), Jost von der Nissen, Nicklass Preußen, and the priest Hans Döbringer are four masters in the KDF tradition who are credited in the Nuremberg Hausbuch with a poem that treats fencing against multiple opponents. Their primary advice is that trying to fight outnumbered is a fool's game, and that even a knightly fencer should not be too proud to avoid it.

According to these authors, the essential problem with trying to fight multiple opponents with a longsword is that a longswordsman only has one weapon. As a result, it is difficult to avoid a situation where a fencer is engaged with one opponent and a second makes an attack. As we will see in Chapter 8, a larger two-handed sword with enough reach can change this calculus. A longsword, however, is necessarily short enough to wear. This places an upper limit on how far at bay a longsword can keep multiple opponents. Since a longswordsman's opponents can stay fairly close to him, it is dangerously easy for them to rush in and attack before the outnumbered fencer can react.

Armizare does not explicitly discuss fencing outnumbered, but Fiore does include a short section on using a longsword against thrown weapons that implicitly reinforces this advice. Fiore sets the scene:

> Here we see three friends who seek to kill this Master, who's waiting for them with his two handed sword ... I wait in this guard ... calmly for them to come at me one after the other, and my defense won't fail against cuts, thrusts, nor any handheld weapon they throw at me.
>
> (Fiore, 2017, p. 31r)

Although Fiore illustrates this scene as a single swordsman being attacked by three men, it is clear that this is a mere pedagogical device (probably to save needing to illustrate three near-identical scenarios). Even in the fictional scenario, as Fiore says, the attackers will not attack together but "one after the other."

While Fiore probably did not intend the above paragraphs as literal advice about fighting outnumbered, the four masters quoted in the Nuremburg Hausbuch would probably agree that the best advice when outnumbered—other than not getting in that situation in the first place—is to fight only one opponent at a time.

OF ROYAL BLOOD   35

Questo magi*r* spetta questi doi cu le lor lame
lo p*ri*mo uol trar cu la punta soprman, E lattro uol

Their essential piece of fighting advice if a fencer must fight multiple opponents is to fight them one after another, engaging only the opponent who is at the end of the line. This is good advice when fencing outnumbered with any armament, but with a relatively short weapon like a longsword it is especially critical. For a longswordsman to engage the center of a line of opponents is to invite the opponents at either end of the line to surround him. As we have already discussed, a longsword is not long enough to keep the surrounding opponents at bay effectively in such a scenario.

Multiple opponents will not, of course, always attack a fencer in a neat line that approaches from one direction. The outnumbered fencer may well have to maneuver until he has gathered all of his opponents on one side of him. There exists a risk during this phase of the fight that the outnumbered fencer will become bogged down fighting one opponent before he can break free of his attackers, and end up surrounded anyway. This is an inherent risk of fighting when outnumbered, which no tactic can completely eliminate. The best a longsword-armed fencer can do in this circumstance is to follow the masters' advice not to stop moving and employ only defensive parries until the opponents are strung out: an outnumbered fencer, in other words, should first employ his weapon to keep himself alive until he has broken free and gathered his opponents on one side of him. Until that objective has been achieved, he should only be parrying, not attempting to strike his opponents.

The masters offer two further technique suggestions for the phase of the fight in which the outnumbered fencer is trying to break free, referred to as the *pfobenzagel* and the *krauthacke*.

Both of these techniques are extraordinarily simple. To execute the pfobenzagel ("peacock tail"), the fencer holds the sword with the point forward and circles the tip, so that the whole sword describes a cone with the wide end toward the opponent (or, as the technique name suggests, a fan-shaped peacock's tail).

The krauthacke ("herb hoe") is executed simply by striking diagonally up from the left or right, then diagonally down along the same line, stepping forward with every rising cut (as if the fencer were hacking at a line of earth with a hoe).

Neither of these techniques is exactly rocket science, but they have a critical common characteristic: they move the sword threateningly without ever actually attacking the opponent. This brings us to a critical psychological asymmetry that drives most medieval advice about fighting outnumbered: when the attackers know that they outnumber their opponent, they are usually reluctant to sacrifice themselves for the good of the many. As a general rule, one attacker of four will back away from an outnumbered defender who is advancing menacingly (with the pfobenzagel or krauthacke, say) rather than risk crossing blades with the victim. After all, while the defender's attention is fixed on him, his compatriots can circle around to attack the defender from the side or behind! Should the attacker being threatened move to engage with the outnumbered fencer, he can significantly increase the odds of a successful "sneak attack" by one of the other attackers, of course, but the attacker also runs the risk that he himself will die.

Exactly how reluctant a set of attackers will be to press their advantage is not something the defender can count on, of course. For this reason, the masters write, an outnumbered fencer may well need to swallow his pride and run before striking any blows.

Indeed, they suggest, the most reliable way for an outnumbered longswordsman to gather all his opponents on one side of him is to run like the wind (with his sword thrown over his back for some protection). If they catch up with him, they will likely be strung out, and the fleeing fencer will have a brief moment in which he can engage one opponent at a time. If he can outrun them, so much better—after all, as the masters' advice repeatedly emphasizes, there is no shame in fleeing when outnumbered.

No fencing master recommends fighting when outnumbered—not even a knight fighting against peasants. However, as the historical advice discussed in this chapter hopefully makes clear, the longsword is especially ill suited to facing multiple opponents. Even a longswordsman who actively maneuvers against his opponents and takes good advantage of the psychology of his opponents is at a significant disadvantage. The four masters of the Nuremberg Hausbuch offer advice for an outnumbered longswordsman to maximize his chances, but that should not be confused with a likelihood of success. Their advice is merely making the best of a dire situation. Even in the best-case scenario, they offer a sobering assessment of the longswordsman's chances: that even the best fencer is usually overcome by multiple opponents in the end.

Fiore calls the longsword "of royal blood" due to its versatility. As we have seen, it is swift, elegant, and highly maneuverable, capable of quick transitions between cut and thrust. Its two-handed grip gives it powerful leverage against other weapons, while its length makes it an excellent grappling aid. All these qualities make it especially useful as a sidearm for armored fighting (for more on which, see Chapter 12). It is no wonder that this was the quintessential knightly sword of the Late Middle Ages.

And yet, for all its virtues, the longsword was not the most popular sidearm of its day. That distinction belongs to the sword and buckler, and it is to this weapon combination that we now turn.

Sidebar: Which would win: longsword or spear?

Spend enough time in HEMA spaces, and I can almost guarantee that someone is going to ask you whether a longsword or spear is the better weapon. The most literal-minded answer to this question—and to all such questions—is that it depends on context. Every medieval weapon was the best choice for at least one situation, however narrow. If it were not, it would not have continued in use. In the case of longsword vs. spear, however, the spear has a number of features that make it the superior weapon in the majority of contexts.

The *best* advantage that a spear has over a longsword is its reach (reach is, almost without exception, the single best advantage one weapon can have against another). But this means more than it may appear at first glance.

People often talk about reach as if they were discussing range on a particularly granular battle map for playing a game, in two dimensions. In actual fact, though, range in fencing is measured in three dimensions, and the reach of a weapon is described by a convex-bottomed, roughly cone-shaped volume whose point is anchored in the wielder, at shoulder height, and whose wide end projects toward the opponent. This cone-shaped reach is longest at shoulder height and shortest when targeting the opponent's foot. If the overall reach differential is great enough, a fencer can hit the opponent's foot while the opponent is still struggling to reach the chest.

In the case of a spear vs. longsword fight, of course, the reach differential is very large. As a result, spears can hit more places from a given distance than can a longsword. It is probably obvious that a spear fencer can target a longsword fencer's head from outside of the longsword's reach. It may not be obvious that the spear fencer can *also* reach the longsword fencer's *foot* from outside of the longsword's reach. This makes the spear's targeting much harder to predict, which in turn makes it very formidable to fence against.

But this is not the only advantage a spear has over a longsword.

Both because they generally weigh more and have a form factor that allows a fencer's whole weight to be placed behind a thrust, the thrust of a spear is far more powerful than the thrust of a longsword. This is especially relevant when facing armored opponents, for the thrust of a spear is much more likely to get stuck in the mail of (and therefore control the movement of) an armored foe.

In addition, because the shaft of a spear creates a large distance between the fulcrum around which the weapon pivots and the point of the weapon, a spear is quicker to redirect than is a longsword (even though, as discussed in this chapter, the longsword is normally considered a very nimble weapon). The percussive blow of a spear packs more force than the percussive blow of a longsword, whether delivered with blade or pommel. The shaft of the spear also permits it to utilize a greater variety of thrusts than can a longsword. A spear can be thrust like a sword, but it can also be thrust like a pool cue. This permits a spear fencer to vary the speed and reach of attacks, which makes it still more unpredictable.

Does the longsword have any advantages? Yes, but perhaps not the ones you're thinking of. The one real advantage a longsword has over a spear is that it can, at least in an unarmored fight and in the right hands, cut you in half, lop off a limb, or split your skull from crown to collarbone. Hard as spear thrusts are, at the end of the day, they're still stab wounds, and a stab wound is only physiologically incapacitating if it hits the brain or spinal column.

We shouldn't make too much of this advantage, though. For one thing, not every fight *is* unarmored, and the cutting capacity of a longsword largely disappears as soon as metal body armor is introduced. For another, while a longsword cut is certainly capable of dismembering a mammal, modern test cutting on carcasses and comparison to the experience of sabreurs in the 18th and 19th centuries strongly suggests that *most* longsword cuts probably did not dismember. Finally, spears themselves often can cut (depending on which kind of spear we're talking about)—but more importantly, a spear fencer can stab a longsword fencer from a range at which the longsword fencer cannot counterattack. If the pain and blood loss of the first spear thrust doesn't make the longswordsman give up, the spear's reach means that the spear fencer can often simply stab again. A modern analogy may be helpful: longswords can also do more damage to the human body than a single pistol bullet, but that doesn't mean you should bring a longsword to a pistol fight. The same principle applies here.

It is about this point that people tend to bring up the question of getting inside the spear's reach. What about getting in close—really close? This is, of course, the advisable course of action for the longsword. However, unless you're fighting in a very, *very* small space, it's perfectly easy to choke up on a spear, or use it like a quarterstaff as a wrestling aid; it is not as if the spear is useless—even at point-blank range. And *getting up close is extraordinarily difficult*. Spears are very easy to retract, so it isn't a simple matter of setting aside the first spear thrust and then rushing in. As the longsword fencer rushes in, the spear fencer (i) can back up, (ii) can pull the spear point back for another thrust, or (iii) do both. And spear thrusts can be delivered wickedly fast, as well; they are not easy to set aside with a sword. Can it be done? Yes, of course. Is it easy? Not at all.

The end result is one that is well known to anybody who has attempted this matchup in sparring. A skilled and canny longswordsman *can* defeat a spearman, but most of the time, the longsword is simply outclassed.

CHAPTER FOUR

Excellent and useful: sword and buckler

> *I will now present to you a combat with the sharp sword, with a large buckler in hand, which will be an excellent thing and very useful for teaching, as well as to one who must inflict some wounds.*
> —Achille Marozzo (Marozzo, 2018, p. 136)

Though the longsword was the late medieval knightly sword *par excellence*, the sword and buckler was the most iconic sword combination throughout the Middle Ages. These were the arms of a serious street fight, as well as the sidearms of a soldier whose armor was not sufficient to make using a longsword advisable.

When I say "sword" in this chapter, I am referring to any single-handed, straight-bladed, double-edged sword. Swords of this sort could have blades anywhere from 30" to 34" long, and tended to weigh about 2.5 pounds. Such swords are typically referred to in HEMA circles as "arming swords," a name that emphasizes their wearability (when a man-at-arms put on his armor, he was said to be "arming"; hence, an "arming sword" is one suitable for donning as part of getting dressed for battle).

Sidebar: Arming swords vs. sideswords

As we discussed in Chapter 2, several of the sources we're going to use in this book date from the Renaissance, even if their traditions themselves have medieval roots. If you know a bit about the history of medieval swords (and you very well may; you probably didn't pick up this book without reason), you may be aware that swords begin to look quite different in the early Renaissance. When most people think of a prototypical medieval sword, they think of something like an arming sword or longsword: a fairly wide double-edged blade on a simple cruciform hilt. But if you're like many people, you may think of the prototypical Renaissance sword as the rapier: a fairly thin blade with a complicated swept hilt that encloses the sword hand in a confection of steel bars, or behind a solid steel "cup." Rapiers can easily have blades longer than 40", and their cutting capacity compared to an arming sword or longsword is substantially reduced. You may be wondering just how much this physical difference in swords affects the applicability of Renaissance treatises to fencing with medieval weapons.

In point of fact, rapier fencing is not so far removed from medieval fencing as is sometimes believed. In this book, however, we will restrict ourselves to Renaissance texts that focus on what HEMA fencers call a "sidesword." This is a one-handed, straight-bladed, double-edged sword with a blade slightly narrower and slightly longer than that of a typical arming sword (anywhere from 30" to 38"). A "typical" sidesword hilt offers slightly more hand protection than the simple cruciform hilt of an arming sword, such as a steel ring to protect a forefinger hooked over the crossguard, a small ring and post that projects at right angles from the blade to protect the thumb and back of the hand, and perhaps a curved bar (a "knucklebow") to protect the fingers. However, sidesword hilts vary significantly in complexity, including simple cruciform hilts that would not have been out of place in the 11th century. Probably for this reason, sidesword treatises do not assume that a fencer has any hand protection beyond a crossguard, making their techniques perfectly applicable to earlier arming swords.

Indeed, much as the division between "Middle Ages" and "Renaissance" is an imposition by later historians on a period characterized by continual change, the division between "arming sword" and "sidesword"

is essentially arbitrary. Is a simple-hilted one-handed sword with a 34" blade a typical sidesword or a slightly longer than average arming sword? The question is nonsensical. Likewise, there is no defensible line to be drawn as to when a sword's hilt is so complex that the sword has ceased to be an arming sword and has become a sidesword.

As with most sword terms, both "arming sword" and "sidesword" are modern terms. In the original German, Italian, and Spanish of our texts, both "arming swords" and "sideswords" were simply called "swords" (*schwert*, *spada*, or *espada*). Even the modern names mean the same thing. An arming sword is a sword that you can arm with; i.e., wear. A sidesword is a sword that you can have at your side; i.e., wear. Thus, even in modern usage, "arming sword" and "sidesword" should be thought of as poles on a spectrum rather than mutually exclusive categories.

The term "buckler" is easy to mistake in English-language discussions of historical fencing. Particularly in older English, "buckler" can be a generic word for "shield," regardless of size. In a modern historical fencing context, it is much more specific than that. As used in this book, a buckler (Italian *brocchiero*, Spanish *broquel*, German *bügkeler* or *pucklär*) is a small shield suitable for wearing on the belt and held by a rigid center grip. Bucklers can be made of wood or metal, and typically weigh about 2 pounds. They can be convex or concave and come in a variety of shapes—typically simple circles, but Talhoffer illustrates an oblong buckler with a crenelated edge, while Kal illustrates several shaped like snarling faces. They also come in a variety of sizes, ranging roughly from 9" to 15" in diameter.

A buckler is considerably less capable than a full-size shield, particularly in an environment like a battlefield where a fencer may be subject to threats from multiple opponents armed with a large variety of weapons (including bows, crossbows, and, by the Late Middle Ages, firearms). However, it is small and light enough to wear on the belt, making it a popular sidearm for soldiers who could not wear a large shield (such as archers), and far more practical to wear in a civilian context for self-defense.

Before we begin a discussion of how the buckler affects technique, we should discuss what makes an arming sword (or sidesword) unique from a fencing perspective. The ability to use a buckler (or other companion weapon; see Chapter 6) *is* one such distinctive, and the one that comes most readily to mind for most people. However, there is another important difference between fencing with a cut-and-thrust sword intended for one hand, like an arming sword, and one intended for two hands, like a longsword: the prevalence of wrist cuts.

By "wrist cut," I mean a cut in which the fencer generates momentum for his cut by rotating the sword in his hand, rather than by rotating the arm at the shoulder or elbow (although "wrist cut" is a conventional system-agnostic term for such cuts in HEMA, in practice, many wrist cuts involve a small amount of elbow movement). Such cuts typically travel a full 360 degrees, wheeling about the hand. Different traditions use different words to refer to wrist cuts, including *schnappen* ("snapping"), *tramazzone* (uncertain), and *mandoble* ("double hand").

Whatever their name, wrist cuts are an important part of fencing with a one-handed medieval sword for a counterintuitive reason: one-handed swords are, effectively, very heavy.

This fact often surprises new fencers. In absolute terms, of course, an arming sword or sidesword is a very light weapon indeed. Two and a half pounds is not a large weight in most contexts. Controlling such a weight at the end of a fully extended arm, however, places strenuous demands on a fencer's hand, wrist, and forearm. Most of my own students who fence with one-handed swords quickly discover that they tire much more quickly using a one-handed sword than they do using a longsword (anecdotally, I have also observed that one-handed swords present a greater risk of tendonitis in new students than do longswords). The reason why is easy to see: a longsword weighing 3.5 pounds is less than double the weight of an arming sword weighing 2.5 pounds, but the longsword is wielded with twice as many hands on the weapon.

The greater relative weight of an arming sword compared to a longsword is exacerbated by the fact that a fencer has two points of contact on the handle of a longsword with which to redirect it, while arming swords are held at only one point on the handle. The upshot of this is that arming swords are slower to attack and slower to recover than a longsword. Wrist cuts are a way to mitigate this problem. A wrist cut does not attempt to stop the momentum of an arming sword in motion but simply to redirect it, so that the sword's own momentum helps to compensate for some of its comparative slowness (please understand that I am speaking in relative terms, here: the cut or thrust of an arming sword is still blindingly fast by most standards; it is simply *less* blindingly fast than the cut or thrust of a longsword). Paradoxically, preserving the sword's momentum is more important with an arming sword than with a longsword, despite the one-handed sword's lesser overall weight. Because of this, medieval one-handed fencing often has a circular flow to it that longsword fencing lacks.

So much for the swords we will discuss in this chapter. What about the bucklers?

Sword and buckler is the sole focus of our oldest extant fencing treatise, the Walpurgis Fechtbuch, which dates to the 1320s. The Walpurgis Fechtbuch is a German treatise that does not appear to be related to any extant fencing tradition, but it presents a well-developed system of stances and counter-stances that strongly suggests it belongs to a tradition now lost. Combined with the age of the Walpurgis Fechtbuch itself, this is a powerful testament to the enduring popularity of sword and buckler fencing in the Middle Ages.

Aside from the Walpurgis Fechtbuch, sword and buckler combat is treated by KDF, Bolognese, escrima comun, and briefly by Monte. In

KDF and escrima comun it is an auxiliary weapon set, treated as a minor variation on the other weapon combinations to which those traditions devote more attention. By contrast, sword and buckler was probably the exemplar weapon of the Bolognese tradition. While dall'Agocchie would insist in 1572 that "understanding is founded upon the unaccompanied sword, and in it one comprehends the entirety of fencing," sword and buckler comprises by far the largest portion of the earlier treatises of Marozzo and Manciolino (dall'Agocchie, 2018, p. 14). Both those authors present only brief material on the unaccompanied sword, but offer multiple extended *assalti*, or forms, for sword and buckler. The length and complexity of these forms suggests that they are traditional teachings from an earlier era of the tradition when the school's focus was on the sword and buckler.

Sidebar: What about the sword alone?

In the fantasy novels of my youth, the protagonist typically fought with an arming sword (often called a "longsword" in the parlance of the times) in one hand, without any other weapons. This seemed to be especially true if the protagonist was a knight or highly born. Following the usage of our treatises, HEMA fencers typically refer to this as fencing with "sword alone" or with "unaccompanied sword" (since the sword is not "accompanied" by a "companion" weapon, such as a buckler or shield).

Fencing with an unaccompanied one-handed sword is a topic to which medieval treatises devote almost no attention. With one important exception (which we'll discuss below), it appears to be a modern invention.

By the 16th century, we begin to see references to the fact that bucklers are not always socially or legally acceptable. We have already seen dall'Agocchie's statement that the unaccompanied sword "comprehends the entirety of fencing." Manciolino says something similar:

> Fencing with sword alone is much more profitable than any other weapon, since you are seldom without your sword. While you do not have your *rotella* or buckler with you at all times, your sword can always be at your side.
>
> (Manciolino, 2010, p. 76)

It is striking that, despite this sentiment, Manciolino's treatise teaches the basics of fencing assuming that the student is using a sword and buckler, and presents no fewer than four forms for that weapon set, as well as a significant amount of additional material not organized into a form. Indeed, unaccompanied sword receives the least of Manciolino's attention.

This perfunctory treatment of a weapon combination that is allegedly "much more profitable than any other weapon" appears to be the result of the changing face of violence during Renaissance Europe. Consider dall'Agocchie's reasoning for preferring the sword alone:

> The principal reasons why the sword is preferred to other arms are that, first, as there is nothing in the world more highly prized than honor, which consists of conducting yourself virtuously, then if someone's were to be placed in doubt through accusation of villainous undertakings or of some shortcoming, he would have to defend it with his own valor, and the other party would have to legitimize his assertion; and one sees how suitable the unaccompanied sword is for doing this. And because those who show themselves to be more resolute in duels appear more courageous and of greater valor, they will appear in a shirt, with an unaccompanied sword, and thereby demonstrate the most manifest proof, putting their faith more in reason and their own virtue than in any other covering or the accompaniment of other arms, either offensive or defensive.
>
> (dall'Agocchie, 2018, p. 14)

Dall'Agocchie's argument is explicitly social: a man must know how to fence with sword alone because that is the most courageous-looking weapon combination with which to defend his honor. A man who insists on being heavily armed and armored for a duel does not show as much confidence in his own cause than a man who dares to fight in nothing but a shirt, carrying nothing but his sword.

This vision of honorable conduct is distinctly un-medieval in at least two ways. For one thing, honorable conduct in avenging an insult for most of the Middle Ages did not necessarily require a one-on-one or even a "fair" fight. Social codes of conduct did apply, but successful vengeance was far more important than modern ideas of fairness. For another, dueling while fully armed and armored was not seen in the Middle Ages as evidence of cowardice. It was, instead, a way to show a fencer's wealth and status. In other words, dall'Agocchie's argument for the utility of the unaccompanied sword is not one that a Bolognese master in the Middle Ages would have been likely to make.

In addition to this, the Renaissance saw changing socio-legal attitudes toward personal violence. In the waning years of the Middle Ages and into the Renaissance, central authorities increasingly inserted themselves into the system of feuds and vendettas that had traditionally governed European concepts of honorable violence. Through a combination of legal measures and social pressures, these nascent forces of law enforcement slowly pushed the traditional concept of honor away from the feud and toward something more recognizable to modern eyes, in which the violence was confined to the aggrieved parties. As the threat of internecine street violence waned, so too did the amount of armament a man might acceptably carry in open society. When Manciolino says, in the 1530s, that a fencer might not always have his buckler to hand, he was speaking literally. A sword was seen as an acceptable self-defense implement; adding a buckler might be gauche or even illegal.

In other words, the importance of the sword alone in the Renaissance is due to a combination of social factors that did not exist during the Middle Ages. No surprise, then, that medieval treatises focus on the much more capable combination of sword and buckler. Fencing with an unaccompanied arming sword must remain the preserve of fantasy.

There is one significant exception to this, which is fencing with the *langes messer*. This was a one-handed single-edged sword of German origin with a T-shaped crossguard whose extra protrusion protected the back of the hand. The name langes messer means "long knife" in German, but the weapon has a blade in the 24" to 30" range, which makes it a sword for all practical fencing purposes (many cultures seem to use the word "knife" to refer to relatively short one-handed chopping swords; cf. the ancient Greek *makhaira* and the Chinese *dao*). The langes messer belonged to a family of similar one-handed cutting weapons of different sizes, ranging from relatively short machete-like weapons to two-handed "kriegsmessers" that rivaled a longsword in size.

The smaller messers can be grouped with the short, single-edged, one-handed cutting swords that can be found across medieval Europe, including the ubiquitous falchion and the Italian *storta*. However, only the messer is discussed in extant treatises (judging by the number of German masters who treated the topic, messer fencing was evidently quite important in the Holy Roman Empire's fencing culture). Most treatises that discuss messer fencing assume that the fencer is armed with an unaccompanied messer. Those that treat the accompanied messer (Talhoffer, for instance, depicts messer and buckler) tend to treat the weapon as essentially interchangeable with the arming sword.

Both these facts make messer fencing an attractive choice for those looking to imagine what unaccompanied arming sword fencing might have looked like. There are two important caveats to this approach. The first is, of course, that there is precious little evidence that fencing with an unaccompanied arming sword was a medieval practice at all. The second is that unaccompanied messer fencing is characterized by heavy use of the free hand to grapple the opponent or his weapon. This is a natural consequence of the fact that messers tend to be shorter than arming swords. The shorter engagement range leads to more grappling opportunities than would be present with longer weapons.

In German and Italian sword and buckler traditions, the principal role of the buckler is to protect the sword hand. In other words, the two weapons are not used as two, but as one. This tradition in German sword and buckler is at least as old as the Walpurgis Fechtbuch (and thus, likely, somewhat older than that actual manuscript). That treatise heavily emphasizes turning the buckler in the hand so that it covers the sword hand, almost enclosing it under a protective half-dome.

Andres Lignitzer, whose teachings on sword and buckler were included in many collections of KDF teachings, makes this point explicit in his first technique:

> Mark when you drive the Oberhaw to the man: with the pommel go inwards, your sword close to the buckler and your thumb, and thrust in from beneath to his face. Wind against his sword and then go with a snap over and around. This works on both sides.
> (Von Danzig, 2010, p. 166)

In this technique, the attacking fencer throws an overhead cut from the right to the opponent. As the sword arm extends, the arc of its cut passes close to the inside of the fencer's buckler, such that the pommel of the sword is close to the thumb of the buckler hand. In this position, the palms of both hands face each other, and the buckler forms a protective shield on the left side of the sword's hilt, laying almost flat against the left forearm. Lignitzer assumes that the opponent parries from his own right, so the attacking fencer's buckler placement protects the hand and forearm should the opponent choose to strike the hand or forearm rather than parrying the blade. The attacking fencer first attempts to out-leverage the opponent's parry by "wind[ing] against his sword," which entails raising the sword to cross further away from his hand and turning the front edge into the opponent's sword to align the hand and wrist so that the opponent is pushing the attacker's sword wrist into the arm rather than trying to bend it. Should this be unsuccessful, however, the attacker can "snap" (a term that indicates a quick, wheeling wrist cut in KDF terminology), executing a wrist cut to strike over own buckler to the right side of the opponent's head.

Important to executing this technique, and many others, is the fact that the buckler is held by a rigid handle gripped in the fist, rather than strapped to the arm. The rigid center grip, together with the buckler's low overall weight, allows the buckler to be rotated in the hand with nothing more than the fingertips of the buckler hand or thumb on the inside face of the buckler. The buckler can be held with its face to the opponent, maximizing its defensive coverage, then flip to the left to accommodate a fencer's sword and protect the hand, as in Lignitzer's play.

The buckler's ability to protect the sword hand is likely an important answer to the question of why European swords did not develop complex hilts until the Renaissance, when bucklers began to be phased out (and even then, older styles of sword persisted). The buckler provides better hand protection than all but the most complex hilts, and covers a significant portion of the forearm as well. A complex hilt actually somewhat hinders the use of the buckler, by forcing sword and buckler further apart, and can even make wrist cuts such as the one we see above harder to execute.

In the Bolognese tradition, the buckler still serves to protect the hand, but it tends to do so through geometry rather than by forming a protective dome directly over the sword hand. For instance, consider the following sequence of attacks from Manciolino. He begins with the sword raised high over the fencer's head, buckler pushed far forward, "making sure that your buckler keeps your head well defended"; i.e., with the buckler pushed far forward and held at roughly shoulder height:

1. Pass forward with your right foot into a wide stance, delivering a mandritto to the opponent's face that goes into Sopra il Braccio, and

come back with a riverso lowering your sword into Coda Lunga e Stretta, making sure that your buckler keeps your head well defended.
2. Pull your right foot near your left, quickly raising a montante to Guardia Alta.
3. Pass forward with your right foot, delivering a fendente into Guardia di Faccia.
4. Pass with the left foot towards the opponent's right side, delivering a stramazzone into Cinghiara Porta di Ferro, guarding your head with your buckler.

(Manciolino, 2010, p. 96)

In this sequence, the fencer makes forehand and backhand cuts at the opponent's face (step 1), throws a rising parry (step 2), and an overhead cut (step 3) that is turned into a wrist cut to avoid a parry (step 4). My point, however, is not the precise sequence of attacks. My point is that throughout the sequence, the attacking fencer's buckler doesn't move. It remains extended from the fencer's left shoulder, "guarding your head."

While Bolognese do mate the sword and buckler hands in the German fashion, they do so comparatively rarely. Instead, Bolognese sword and buckler fencing is replete with admonitions to keep the buckler pushed as far forward as possible. This has the function of allowing a small shield to obscure a large area. Think about how you can cover the sun with a quarter. The small object can obscure the large object because the quarter is very close to you and very far away from the sun. Bolognese buckler use operates on the same principle. By pushing the buckler forward, a Bolognese fencer obscures her sword hand (and a significant part of her own body) from the opponent's perspective while permitting the sword hand largely unrestricted motion.

One consequence of this is that Bolognese sword and buckler fencers rarely parry with the buckler. Paulus Kal's very first sword and buckler play uses the buckler to parry a high cut while the sword strikes low. Monte, too, recommends learning how to fence with sword and buckler (which he refers to in Latin as a *pelta*) first by learning to parry with the buckler alone, without any sword at all:

> To play with sword and *pelta* at first it is good to teach ourselves to cover without sword, with only a *pelta*, as we do with the sword when we have no *pelta* since otherwise a long time would pass before we would know correctly how to cover ourselves.
>
> (Monte, 2018)

Bolognese buckler use, on the other hand, is largely content to let the buckler defend the fencer's upper left quadrant more or less passively, incentivizing the opponent to attack elsewhere (and, thus, making his attacks more predictable). We will see a similar ethos predominate in Bolognese sword and shield usage in Chapter 5.

In escrima comun, on the other hand, the buckler is free to move to any quadrant of the body and largely disconnected from the sword. Godinho, our only surviving source for escrima comun, begins with the buckler held in a fashion that would be familiar to any Bolognese fencer:

> He that will give battle with sword and buckler is obliged to hold the buckler arm long and straight, because if he holds it withdrawn, the elbow remains uncovered, and going straight, it cannot be injured.
>
> (Godinho, 2016, p. 71)

However, he goes on to advise that the buckler be moved however necessary to proactively intercept any attack:

> When taking the parries, he doesn't wait for the opponent's sword to come to it [i.e., the buckler]; he is obliged to halt the blow in its path. Stated more clearly, the buckler-man is obliged to seek the opponent's sword when it comes with the blow, because it is so small, one can be swiftly deceived in the parry.
>
> (Godinho, 2016, p. 71)

Just as Bolognese buckler work is very reminiscent of how Bolognese uses larger shields, this active use of the buckler is also characteristic of how escrima comun freely uses moves, even a large shield to a fencer's left or right, high or low, to defend against attacks. We will see more of this in Chapter 5. Indeed, Godinho speaks of the buckler almost disdainfully as a mere derivative of the larger *rodela* shield (Godinho has previously clarified that while large shields intercept attacks on their face, bucklers do so with the rim):

> Those who have lost interest understand that whoever plays well with the shield (as it is queen) will be excused from learning dagger or buckler, as he can use the same play in whichever of said

weapons, save that one differentiates the placement of the buckler and taking of the parries.

(Godinho, 2016, p. 71).

Lignitzer's fifth play offers an example of how the buckler can be easily deceived if used in a static manner. In Chapter 3, we saw that one of the major distinctives of medieval German swordsmanship was the use of the sword's back edge to attack along horizontal or descending angles. Lignitzer's fifth play begins with such a cut, a plunging stroke down from the right side using the back edge called a *sturzhau* (*sturz* meaning "plunging") that he turns into a thrust:

> From the Sturzthaw (plunging strike) make as if to go to [his] left side over his shield with a thrust then with the point change under and thrust swiftly inside his shield. Wind immediately to your left side and if he defends against this then take his right leg with your long edge.
>
> (Von Danzig, 2010, p. 166)

The play begins with the fencer cutting from above on the right with the back edge. The palm of the fencer's sword hand will be facing to the right as the cut plunges in point first, almost like a thrust. This has the effect of throwing his point in at a high angle, over and behind the opponent's buckler. This attack is easily defeated by raising the buckler: the attacking fencer's point is already coming in at such a high angle that the attacker cannot likely raise it further to angle around a parry. What the attacker can do, with a simple twitch of the wrist, is dip the point down and to the left, snaking around the defending buckler to thrust inside, between the opponent's buckler and sword. In so doing, the attacker "winds" his sword by rotating the sword hand so the palm of the sword hand is facing up. This gives the attacker the freedom of motion to wind the sword point further to the right if the defender's buckler tries to parry the thrust by pushing the sword to the defender's right. The defender must instead parry with the inside thrust with the sword.

Following Lignitzer's earlier advice, the attacking fencer will keep the buckler close to the sword hand during these actions, keeping it between the sword hand and forearm and the opponent's blade. However, when the defender tries to parry the inside thrust with the defender's sword, the attacker can keep the buckler high, pressed against the opponent's sword hand, leaving the attacker's own sword free to cut the opponent's leg.

An escrima comun fencer, analyzing the defender's actions in this play, would probably criticize the defender's initial action. From the escrima comun perspective, by waiting to simply nudge the incoming attack aside with the buckler, the defending fencer was open to the sword dipping under and around. Instead, the defender should have moved forward with the buckler parry, "seek[ing] the opponent's sword," and thereby stifling the attack.

The opening of Lignitzer's fifth play also raises an opportunity to discuss buckler shape. One of the reasons that the attacking fencer can circle the thrust down and around so easily is that the sword is following the curve of his opponent's buckler. However, not all bucklers were round.

The typical medieval buckler was round, of about 9" to 15" in diameter. Indeed, Monte uses the buckler's shape to describe the hand guard of a warhammer, advising that a good warhammer should have a disc-shaped handguard. However, Talhoffer illustrates an oblong buckler with a scalloped edge, and both Marozzo and Manciolino discuss the use of a small center-gripped rectangular shield with a wavy face called a *targa* (not to be confused with the Scottish *targe*, an arm-strapped round shield).

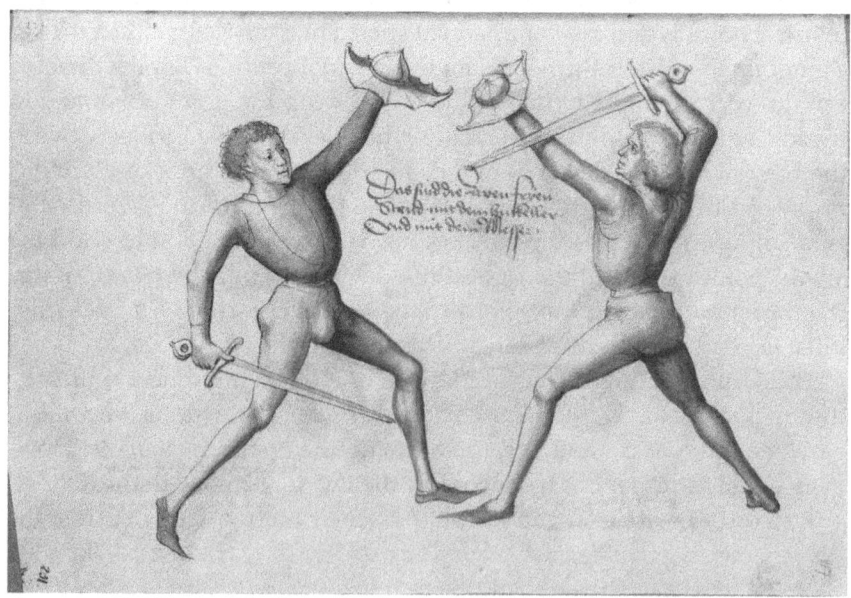

These non-round arms are typically considered a species of buckler in modern HEMA parlance, and indeed are used in largely the same way. However, these unorthodox shapes do provide some advantages that round bucklers do not. One such advantage is the ability to deflect thrusts with a simple turn of the shield. Consider Marozzo's eighth technique from his first form with sword and targa, defending against an underhand thrust (stoccata):

> Let's first propose that he wants to throw the stoccata to your face. Beat this stoccata to the outside with your targa, and stepping forward with your right foot, throw a mandritto to his legs or a thrust into his left flank.
>
> (Marozzo, 2018, p. 154)

It is interesting to note that nowhere does Marozzo advocate beating a thrust to the outside with a round buckler. Inexperienced fencers are often tempted to do so, but, as Lignitzer shows, it is easy for a blade to slide around the rounded lower edge of a circular buckler. A defender with a targa, however, actually can parry a thrust by simply rolling the hand over. This causes the targa to rotate, and the corners of the targa catch and flick the incoming thrust away. This allows the targa-armed fencer to defeat a thrust with a very simple motion, one which does not even move the targa from its overall location in space. Talhoffer's scalloped bucklers can be used in a similar manner.

The tradeoff for this extra defensive capability is that a targa or other unconventionally shaped buckler requires more material to cover the same area as an equivalent round buckler. The difference is usually slight in absolute terms, but a fencer can immediately feel the variation of even a few ounces when a buckler is held at the end of extended arms or weighing on the sword belt. A larger or unconventionally shaped buckler may be a more capable weapon, but it is also less convenient to carry and more tiring to use.

In short, medieval fencers had to balance shape, size, and weight when choosing a buckler. Both Marozzo and Manciolino discuss this in the context of training. Their most complicated sword and buckler material is explicitly described as for the sword and "small" buckler. Forms for the sword and targa and sword or "large" buckler are simpler, but explicitly noted as intended for use with a sharp sword. The implication seems to be that students should learn to fence with a small buckler, but when fighting in earnest, rely on less flashy techniques and more capable bucklers. After all, a fencer may need to practice for long hours, but an earnest fight is usually over far quicker!

Speaking of earnest combat, any discussion of earnest combat with sword and buckler invites questions about offensive use of the buckler. Thus far, we have discussed the buckler only as a defensive implement. Hold one in the hand, however, and it is easy to see why fencing masters described it as a weapon. Even completely untrained members of the public who pick up bucklers at public demonstrations often make punching motions with them. A weighty disc of rigid material gripped in the fist fairly cries out to be punched into things.

Bucklers certainly do significantly enhance a fencer's punches, and there are scattered references to buckler punches in our extant material. In a modern context, it is well understood that buckler punches can land with such force that they constitute a safety hazard. Whether delivered with the boss of the buckler or the rim, a buckler punch can stave in a fencing mask. For this reason, when modern historical fencers fence in "unarmored" safety equipment, it is conventional to require fencers to "pull" their buckler punches for safety reasons, and in some events they may be banned altogether. No wonder that Talhoffer shows their use in unarmored duels to the death.

EXCELLENT AND USEFUL 57

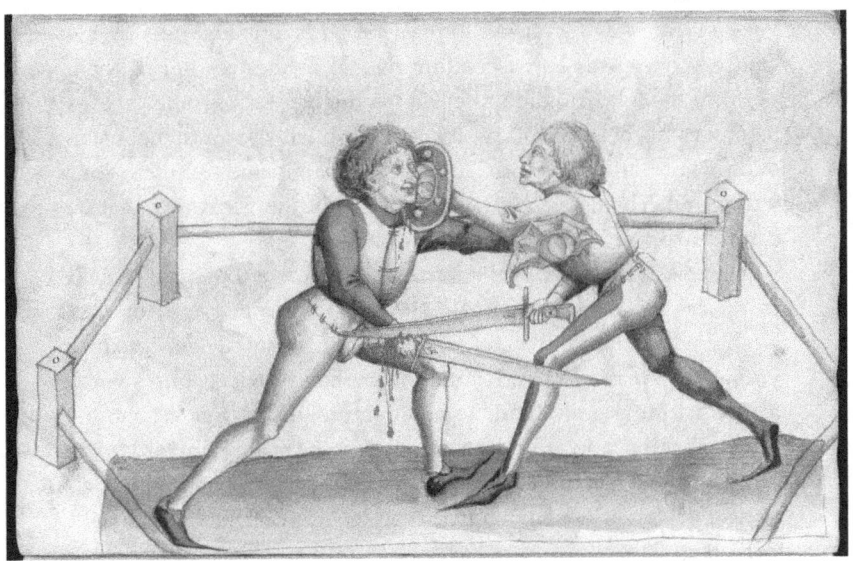

Notwithstanding the bone-crunching impact of a buckler punch, the majority of offensive buckler maneuvers do not involve punching at all. To understand why, we must remember that modern sparring might ban full force strikes with a buckler, but it bans *all* strikes with sharp swords. As we saw with pommel strikes in Chapter 3, it is easy to forget that the blunt implements in sparring (such as bucklers) are real, while the sharp implements (such as swords) are not. Eye-watering as a buckler punch can be, it is rarely as deadly as the cut or thrust of a live blade.

It should not surprise us, therefore, that the majority of offensive buckler techniques in our extant sources involve either grappling or pushing with the buckler to allow a fencer to strike with the sword.

Grappling with the buckler may seem counterintuitive. After all, the grappling fencer's left hand is busy gripping the buckler handle, and thus cannot actually grab the opponent. Nevertheless, grappling is an important part of sword and buckler fencing. Because the buckler hand is occupied, buckler grappling involves using the whole left arm to wrap around an opponent's arm (typically, the sword arm). Following is a good example of a buckler grapple from Manciolino, which also highlights one of the ways in which the buckler enables grapples that would actually be more dangerous to attempt with an unencumbered hand.

> Pass forward with your left foot towards his left side, into a wide stance, while cutting an ascending riverso to his sword-arm. Then, perform the following grapple. Feint a buckler-strike to his face; as he moves his head from fear, thrust your buckler-arm to the inside of his sword-arm and bind it, pulling it strongly and tightly to your left armpit. Then, pass back with your right foot, ensuring that he cannot strike you with his buckler.
>
> Here is the counter. As he passes with the left foot as described to attack you with a riverso, extend your sword-hand forward (well covered by your buckler); as he feints the buckler-strike to your face, further extend your sword-hand (with your sword) forward to invite the bind; as he attempts to grab you, grasp his arm as it comes towards you, strongly pressing it downwards. The opponent will feel the pain from your clutching him and will let go of his buckler; at which point you can leisurely strike his face with a riverso.
>
> (Manciolino, 2010, p. 115)

In this play, after a backhand ascending cut, the attacking fencer feints a buckler punch to the opponent's face, and uses the momentary distraction gained thereby to wrap the buckler arm over and under the opponent's sword arm, pinning it to the attacker's side. The step backward at the end of the play pulls the opponent off balance, leaving the opponent open to the attacker's sword, in addition to avoiding a counter-strike from the opponent's free buckler. Paulus Kal's final sword and buckler technique is a similar arm wrap, except that Kal winds the buckler under and over the opponent's arm and uses the leverage gained thereby to throw the opponent rather than pull.

It is interesting to note that the buckler punch in this play is a mere feint. This is a good example of the relative value historical fencers placed on buckler strikes compared to strikes from sharp blades. Manciolino clearly prefers the more complicated motion of pinning the opponent's sword arm to the simpler act of striking the opponent in the face with a buckler … which, of course, would leave the opponent's sword free to retaliate. In practice, the attacking fencer's buckler may strike a glancing blow in the course of wrapping the arm (especially if the fencer is using a larger buckler). This can be a happy accident, but it is not the focus of the technique for the same reasons that Manciolino prefers to trap the sword arm rather than execute a buckler punch in the first place.

At the same time, it is important not to undersell the effectiveness of a buckler punch to the face. Notice the way Manciolino describes a potential counter. The attacker's arm wrap is a two-part motion: first, a feinted punch to the face, and second, the arm wrap. In response to the feinted punch, Manciolino has the defender counter by extending the sword arm. This pushes the buckler punch away from the face, just in case the punch is not a feint. However, this also makes the defender's sword arm a tempting grapple target, and requires the defender to have a counter-grapple ready (in this case, wrenching downwards as the attacker begins the arm wrap, so that the attacker's wrist is pinned painfully between the defender's sword hilt and the buckler grip).

While a counter-grapple is certainly possible, it is risky. This is good from the attacker's standpoint: the defender must risk either being punched in the face by a buckler or be forced to grapple. If the attacking fencer were not attempting to grapple with a buckler—if the attack was a feinted punch to the face with a bare hand followed by seizing and grappling the sword arm—then none of this would be necessary. In that case, the defending fencer could simply counter the initial punch by cutting the attacker's bare hand, and have done with it. The attacker's buckler prevents this, significantly complicating the defender's potential responses.

Of course, the buckler also means that an attacker must get quite close to grapple—close enough to execute an arm wrap. If the attacker cannot get that close, the buckler can still push on the defender's sword hand, trapping it behind the broad face of the buckler and leaving the opponent open. The following is a good example of a push with the buckler hand, from the first part of Marozzo's first form for sword and targa:

> [S]tick the point of your sword into the enemy's moustache, with your sword accompanied by your targa in guardia di faccia, and immediately advance a big step with your left foot toward the enemy's right. As you take this step, extend your arms, that is, put your targa into his sword hand and throw a thrust into his chest at the same time.
>
> (Marozzo, 2018, p. 151)

In this play, the attacking fencer throws an initial thrust, which the opponent defends (notice that this is one of the cases in which a Bolognese master advocates keeping the hands together so the targa covers the sword hand). As the defender parries, the attacker circles to the left (toward the opponent's sword side) and smashes the targa into the opponent's sword hand. This pushes the opponent's sword hand into the belly or chest and leaves the attacker free to throw a second thrust to the opponent's chest.

Buckler pushes such as this are distinguished from true "punches" primarily through intent. A buckler push can be quite forceful, but its objective is not to cause injury. In the play we just examined, the opponent cannot reach the incoming thrust with the opponent's own buckler, and the opponent's sword is momentarily pinned by the attacker's targa. Whether the attacker "punches" the targa or merely "puts" it into the opponent's sword hand is immaterial, so long as it pins the opponent's hand so that it cannot defend against the second thrust. That said, one of the advantages of a buckler over a bare hand for this type of movement is the fact that the buckler wraps the fencer's "pushing" hand in hard, inexorable steel or wood. When fencing in a serious encounter, the line between a "push" to the opponent's hand and a bruising "punch" to the wrist is blurred.

Whether punching or pushing, a sword and buckler fencer must always be mindful of opportunities to bring his buckler into play as a separate weapon at close range. Often these opportunities are implicit, rather than explicit, in our surviving material. Consider Lignitzer's fourth play:

> From the Mittelhaw (middle strike) make the Twer (cross strike) to both sides and the Schaitlar (skull strike) with the long edge, then make a thrust in underneath.
> (Von Danzig, 2010, p. 166)

Here, the fencer uses the characteristic German zwerchau (see Chapter 3 for a more thorough description of this cut) to strike horizontally to the opponent's left and right sides in quick succession. Following Lignitzer's normal procedure, the sword whips left and right under the protective dome of the attacking fencer's buckler. If neither of these strikes hits, the attacking fencer breaks the pattern by striking vertically to the crown of the opponent's head (again, accompanied by the buckler, serving as a detachable handguard). This pulls the opponent's sword upward into a parry. At this point, Lignitzer's written instruction is simply to pull the sword back and stab the opponent below. However, this final movement is far more effective in practice if the attacker's buckler remains high, pressing the defender's hands up and back so they cannot respond to the belly stab.

The play above is an example of the advantages of having a detachable handguard (i.e., a buckler) rather than an integrated one (such as a basket hilt that fully encloses the hand in a cage of metal). The attacking fencer can keep the sword hand protected for most of the play, but when a close-range opportunity presents itself, the attacker can disable the defender's weapon with bruising force. Were the attacker's hand protected by a basket hilt (which we normally think of as a more "advanced" hilt type), the attacker would have to push upward with an unprotected left hand. Compared to a buckler, an empty hand presents a smaller surface to push into the opponent's weapon, leaves the pushing hand exposed to injury, and cannot strike with as much incidental force. In this way, the buckler offers versatility that an integral handguard does not.

This forceful push with the buckler at close range is a hallmark of the Walpurgis Fechtbuch, which calls it a *schiltschlac* or "shield strike" ("shield" here meaning buckler, as the treatise's illustrations make plain). However, the buckler push can be found in sword and buckler fencing throughout the rest of the Middle Ages and into the Renaissance. The opportunistic use of the buckler as a detachable handguard for so long may be one of the reasons that one-handed European swords retained their simple hilts, even in an unarmored or "civilian" context, when sword smiths had the theoretical capability to craft more complex hilts.

The last topic in sword and buckler fencing we should discuss is fencing against multiple opponents. In Chapter 3, we saw that longswords are ill suited to engaging multiple opponents at once. The fundamental logic of that scenario applies to a sword and buckler as well. A sword and buckler, like a longsword, cannot sweep a very large volume of space around the defending fencer. The likely result is that the attacking fencers will lurk just beyond the sword's range and one or more will rush in to strike the outnumbered fencer when the defender's threatening blade has swept past them. The essential advice for fencing multiple opponents when armed with a sword and buckler is the same as with a longsword: run away rather than engage in such a fight. If running is not possible, the next best advice is maneuver so that the multiple opponents can be strung out and engaged one-on-one. These two pieces of advice are not so different from each other. After all, running away can force multiple opponents to string themselves out.

However, there are two additional pieces of historical information we can bring to the question of multiple opponents with sword and buckler. The first is that Godinho discusses specific cutting patterns useful for fencing with a one-handed sword when outnumbered. The second is that Talhoffer discusses the specific case of how to employ sword and buckler when fencing outnumbered two to one.

We shall take these in reverse order. Talhoffer's two-on-one instructions are exceedingly brief (in truth, they barely merit the term "instructions"). In two of his manuscripts, Talhoffer presents the scenario of a fencer armed with sword, buckler, and dagger (the dagger held, somewhat awkwardly, in the buckler hand) facing two opponents, one on either side. In the 1459 manuscript one of the attackers is armed only with a dagger; the other opponents are all equipped with sword and buckler. In both manuscripts, the surrounded fencer engages one opponent with his sword alone while fending off the other with his buckler and dagger with a caption indicating that this split-weapon stance is necessary in the "emergency" (*notstand*) of two against one. Although the scene is illustrated in a dueling yard, the fact that one attacker is illustrated with a simple dagger, and the defending fencer's unusual armament of buckler and dagger in a single hand all suggest that Talhoffer has a spontaneous self-defense scenario in mind.

Talhoffer's "emergency" stance separates the sword and buckler. This is a radical departure from typical sword and buckler usage, presumably justified by the fact that otherwise the defending fencer would be surrounded, and should not be taken as advice that a sword and buckler fencer should *remain* surrounded simply because he has two weapons. Escaping the situation or maneuvering until the outnumbered fencer can engage a single opponent is still the most practical advice. However, the buckler allows a fencer to make it more difficult for a second opponent to sneak up and land a blow while the defender is engaging the other attacker.

Simply sticking a buckler out behind you does not, of course, present a significant obstacle. Talhoffer's 1467 manuscript provides some additional color, offering this instruction to the surrounded fencer:

> There he displaces with turned hand and becomes himself turned and hews to the behind.
>
> (Talhoffer, 2019)

The surrounded sword and buckler fencer is pictured parrying one opponent's attack with his sword in "turned hand," preparing for a quick wrist cut. With the first attacker's blow parried, the surrounded fencer pivots 180 degrees, redirecting his sword with a wrist cut to threaten the second opponent while thrusting the buckler and dagger at the first. The instructions are brief, but the picture is clear: the surrounded fencer is constantly in motion, turning from one opponent to the next, not engaging in an extended exchange of swordplay with one opponent while blindly trusting his extended buckler to protect his rear. The sword in particular is always in motion, taking advantage of a one-handed sword's ability to preserve and redirect momentum through the liberal use of wheeling wrist cuts.

This constant threatening of one opponent and then the next is also the core of Godinho's advice for keeping multiple opponents at bay with a one-handed sword. In the quote that follows, Godinho describes a basic cutting form together with modification for multiple opponents:

> Putting in the right foot, he makes a *tajo*, extending as much as he is able, and when the *tajo* withdraws, he withdraws the foot that he had placed at the same time, returning it to where it began. Then he returns to put in a *reves*, extending it as much as he is able, and

as the *reves* passes, he withdraws the foot like he withdrew for the *tajo*. Then he puts in a *tajo* with the left foot. If there is more than one opponent, then when he puts in the left foot, it should not be straight from the other, but to the left side so that it reaches more. Then in this form, he puts in the right foot with a *reves*, and cuts with this step as much as necessary, taking note that the eyes should always go where the blow goes. Being necessary to retreat, he should leave by placing the feet with the same step, giving a nails-up thrust to parry those on the left side, and a nails-down thrust to those on the right side, not making a circle with his sword when he turns the sword from nails-up to nails-down, only turning with the wrist.

(Godinho, 2016, p. 4–5)

In the basic form, Godinho has the fencer throw alternating forehand and backhand cuts, stepping forward with each blow. If there are multiple opponents, however, the steps carry the fencer first to the left and then to the right, so that the fencer is threatening all comers. The basic picture is the same as Talhoffer's, attacking first one opponent and then the next in quick succession. In this section of his text, Godinho assumes that the fencer is armed with a sword alone, but it would not be much of an extrapolation for a fencer trained in the escrima comun tradition to use his buckler to intercept blows on one side of his body while using his sword on the other. After all, as we have already discussed, escrima comun is perfectly happy for the sword and buckler to move independently of each other even when fencing a single opponent.

The above rule (as Godinho would call it) has the fencer stepping from side to side but still threatening opponents to the front. It thus works best when the multiple opponents are fairly close together, forming a line in front of the fencer. This might well be the case in a narrow side street, for instance, where there simply isn't room for the attackers to spread out much. Godinho offers a second "rule" for threatening multiple opponents "if one is attacked in a wide street," or other scenario in which the attackers have more room to spread out.

> If one is attacked in a wide street, or unable to reach all the opponents with the blows pointed out in the previous rule, he will put his right foot toward those on the right side and throw a *tajo*, which will cut until reaching those that are on the left side. At the

> same time that the *tajo* cuts, he lifts the right foot and puts it in front of the left foot. Then, he puts the left foot to the left side, giving a *reves* to those on the same side, which will cut until reaching those on the right side....
>
> Note that when you finish the *tajo*, and you want to arm the *reves*, incline the point nails-up at those remaining to the right side, which is not in order to injure, as much as for the parry of the blows that the opponents may throw in the time that the *tajo* passes. Time isn't lost because in order to give the *reves*, one ends with the point already armed.
>
> When you finish the *reves*, and you want to give the *tajo*, you have to turn the point nails-down to those that remain to the left side, not so that you injure with it, but for parrying the blows that the opponents will give as soon as the *reves* passes, and at the same time, the *tajo* is armed with the said point. With this step, you will make the blows that will be necessary, taking note that the body should always go straight, where the blow goes, with your eyes on the opponents.
>
> (Godinho, 2016, p. 5)

This time, rather than stepping right and cutting backhand, the fencer steps right with forehand cuts and left with backhand cuts. This allows the fencer to reach further with each cut, maximizing the volume of space that the sword sweeps and thus maximizing the distance at which the opponents are kept at bay. At the same time, Godinho recognizes that savvy opponents will stay just out of reach of these cuts, waiting until the sword has whistled past them and then springing in with their own attacks. To deter this behavior (and to parry the blows that are surely coming), Godinho inserts a thrust between each cut. The final pattern looks something like this:

1. Step right with a wide forehand cut toward the right.
2. Recover the right foot and shift weight toward the left, thrusting back to the right.
3. Step left with a wide backhand cut toward the right.
4. Recover the left foot and shift weight toward the right, thrusting back to the left.
5. Repeat.

It is worth pointing out that this pattern is not meant to be rigidly adhered to, executed like an automaton. Indeed, as Godinho notes, it is important for the fencer to keep his eyes on the opponents to either side, so that he has a continuously updated idea of where his opponents are and what they are doing. In this way, he can subtly modify the placement of his steps, the trajectory of his cuts, the timing of his thrusts, and other such nuances to best respond to his opponents' movements and intentions. When executed intelligently, the pattern described by the rule can allow a fencer to cover quite a lot of ground, keeping opponents on either side off balance by thrusting back at them just as they think they have a window of opportunity to attack.

The sword and buckler was a powerful and compact weapon combination. Though it has a comparatively small footprint in modern pop culture, in many ways it was the quintessential medieval sword form. In this chapter we have looked at the momentum-preserving quick-wheeling flow that characterizes one-handed medieval fencing. We have seen how the buckler can provide superb protection for both hands, yet it is also a versatile offensive implement with which to control an opponent's weapon or hand—or, in rare cases, simply land a hard punch to the face. We have also looked at some similarities between the use of the buckler and the use of larger shields in both the Bolognese and escrima comun tradition, a topic to which we will now turn.

Sidebar: Which would win: sword and buckler or longsword?

Sword and buckler and longsword are the two most iconic medieval sword combinations, and together comprise the largest portion of our surviving material on medieval swordplay. It is perhaps surprising, then, that no treatise discusses the use of one against the other. It may be that class differences underlie this otherwise curious omission: the longsword was largely a "knightly" sidearm, while the sword and buckler seems to have been associated with fencers of lesser status (though not, we should be careful to note, lesser skill).

In an armored context, the differences between sword and buckler and longsword are stark. As we will discuss in greater detail in Chapter 12, defeating medieval armor with a sword is extremely difficult without being able to thrust with one hand grabbing the sword blade to stiffen it. This is difficult to do safely with a left hand that is also holding a buckler, and a longsword's slightly greater size tends to make it a better wrestling aid against armored opponents. Indeed, the longsword's superiority as an anti-armor sidearm goes far to explain why it became the quintessential knightly sidearm in the Late Middle Ages.

In an unarmored context (strange as such a match might be from a social standpoint) the two weapons are much more evenly matched. The longsword has no significant reach advantage over an arming sword (recall our discussion of the effect of a two-handed grip on reach in Chapter 3), but it does have a significant advantage in speed and leverage. Against these advantages, the sword and buckler can more easily reach a greater variety of targets (such as legs) and can use the buckler to suppress the longsword while the arming sword works.

As we have seen in this chapter, attempting to block an attack with the buckler alone, while a valid historical technique, was acknowledged even by historical authors as a fraught endeavor. This danger only increases against a longsword, with its superior ability to redirect its attacks compared to that of an arming sword (I will also add that longswords hit hard enough that I personally find it physically uncomfortable to block their cuts with a buckler). A longsword's superior leverage also makes it difficult to parry with the arming sword alone, while the reverse is not true. The buckler can help here: an arming sword and buckler held closely together can resist force as well as a longsword can. As a result of these considerations, it is especially dangerous for a sword and buckler to separate too early against a longsword. By the same token, a longsword-armed fencer who can trick or provoke the opponent into separating sword from buckler can often find a way to snake a cut or thrust in.

Provided the sword and buckler fencer avoids this mistake, however, the sword and buckler fencer can neutralize most of the longsword's advantages. The reverse is harder to do. At close ranges (again, provided both fencers are unarmored), a longsword provides poor defense against the sword and buckler's ability to separate and strike both high and low targets (in modern sparring, longsword fencers who try this matchup frequently find their ribs and legs covered in bruises when the arming sword makes a strike that would have been radically unsafe for a longsword to make). The longsword's only reliable defense against this is to keep the distance of the fight open. While this strategy can work, it is inherently risky to have to retreat from close-range fencing, because follow-up attacks typically require fencers to remain briefly at close range. This dilemma calls to mind Manciolino's derisive comment about fencers who can only fence at *gioco largo*, or long range, and cannot fence at *gioco stretto*, or close range:

> If you were only skilled in the *gioco largo*, and found yourself in the *stretto*, you would be compelled with shame and danger to pull back, thus often relinquishing victory to your opponent—or at least betraying your lack of half-swording skills to those who watch.
> (Manciolino, 2010, p. 77)

In an armored contest between longsword and sword and buckler, there is no contest: the longsword's ability to offend an armored man is greater than any other sidearm sword. In an unarmored contest, while the issue is frequently close, a carefully handled sword and buckler has greater ability to match the longsword's advantages than the reverse.

CHAPTER FIVE

Queen of swords: sword and shield

> *It is queen.*
> —Domingo Luis Godinho (Godinho, 2016, p. 71)

When I was young, I was under the impression that shields were rather useless. I was prepared to grant that they had a certain aesthetic solidity, but as I could plainly see from my roleplaying games, they offered very little in the way of actual protection (about as good as light armor, so I thought). Why anybody would give up a two-handed weapon just to wield a shield was quite beyond me. As a historical fencer, my opinion of the shield has changed. As Godinho bluntly declares, "it is queen" (Godinho, 2016, p. 71).

A good many people remain confused and curious about the use of sword and shield. This is a weapon combination that evokes martial traditions across millennia of history, in Europe and beyond. It is the chosen armament of warriors in many a fantasy story (and more than a few in science fiction), and is a staple of fantasy and historical games. And yet, for all this, the technical *use* of the shield is maddeningly elusive.

We should stop here and define terms. The English word "shield" can refer to a truly enormous variety of objects, from tiny bucklers that are barely large enough to enclose the fist to great mantlets large enough to protect multiple men and so heavy that they were mounted on wheels. For purposes of this chapter, however, I shall use the term "shield" to refer to defensive weapons light enough to be used by a single person and large enough that they are roughly no less than 18" in their smallest dimension. This definition excludes the largest and heaviest shields, which operated more like (semi-)portable cover than personal arms. It also excludes the smallest shields, which we discussed in Chapter 4.

Surviving medieval sources that discuss shields of this sort—what we might call "full size" shields—are not very numerous. Monte, Manciolino, Marozzo, and Godinho all include sections on fencing with the sword and *rotella*, a round and slightly domed, arm-strapped shield roughly 2 feet in diameter. The rotella could be made of wood, but also—unusually for a large shield—could be made entirely of steel. Marozzo also briefly discusses the use of the *imbracciatura*, a concave, oblong, arm-strapped shield roughly as tall as the distance from foot to throat with a spike at the lower end (an example of what some people call a "tower shield"). Lastly, KDF and other German sources discuss fencing with a man-height spiked shield gripped by a central handle that runs the length of the shield, which appears to be used exclusively in the context of ritual duels. In terms of substantive technical fencing information on the use of the sword and shield, that is it.

Missing from this list is the center-gripped round shield that dominated Norse warfare, the oblong "kite" shield that proliferates across Norman warfare, and the triangular "heater" shield of the High Middle Ages. Collectively, these three forms of shield dominate much of

Sidebar: Arm-strapped or center-gripped?

We can spare a few words to sketch the broad strokes of the differences between the arm-strapped shields for which we have better evidence and center-gripped shields such as the iconic Norse or "Viking" shield. With the caveat that the use of the center-gripped shield is essentially all speculative, we can tentatively make this generalization: that arm-strapped shields are better armor, and center-gripped shields are better weapons.

With a center-gripped shield, about half the length of the shield projects past the hand, and the entire shield can rotate around the hand. As a result, the shield can act as a door. For instance, suppose a fencer strikes a descending diagonal blow from the right, at the head of an opponent armed with a center-gripped shield. The defender can intercept with the shield while moving forward and to the defender's left. As the defender moves to the attacker's right, the defender can move the shield arm across the body and let it rotate in the hand to flatten the opponent's sword arm into the opponent's body like a door, and then strike the knee or the head, below or above the shield.

Experimentally speaking, a lot of offensive work with center-gripped shields seems to work out that way, or variations of it. The center grip places the shield further out on the arm, and allows it to rotate in the hand. Thus, it's easier to make it a wall on either side of the body, opening up more possibilities to press the opponent's weapons into the body or otherwise get around them. What little technical information we have on center-gripped shields seems to corroborate this. For example:

Experiment also suggests that the projection of the shield beyond the hand of a center-gripped shield is quite useful for punching people in the face and hooking the opponent's shield. These are plausible-seeming uses for a shield, although as always, we come back to evidence problems when we try to prove that they were definitely historical techniques.

At the same time, the ability to rotate the shield in the hand and the length of shield projecting beyond the center grip can be a liability. If the opponent can press on the part of the shield that projects beyond the hand, the shield will rotate inward, which can rob a fencer of the shield's protection and foul up the weapon arm into the bargain. In this image from Talhoffer's 1459 treatise, the left hand fencer is opening his opponent's longshield in this way by using his foot, but with a round shield there are equivalent techniques.

In short, just as center-gripped shield users can take advantage of the shield's ability to rotate in the hand, so they must always beware of their opponents taking advantage of the shield's ability to rotate in the hand.

Arm-strapped shields do not present this weakness, which is certainly one plausible reason as to why fencers might choose one style of shield or the other. An opponent has a much harder time manipulating a shield if it's strapped solidly to the arm, because such shields do not rotate around the hand. Manipulating an arm-strapped shield is not impossible (a strong press can still collapse the shield arm against the chest), but it is definitely harder. Arm-strapped shields also allow a fencer to use the shield hand to hold things, such as the rungs of a siege ladder or the haft of a spear, by the simple expedient of sliding the hand through the straps past the rim of the shield. Center-gripped shields have no such analogous capability.

At the same time, because it can't rotate in the hand, it is much harder to use an arm-strapped shield to suppress the opponent's weapons (though it is not impossible; we'll look at an example later in this chapter). This difficulty is compounded by the fact that the shield does not project very far beyond the hand, making it less useful for hooking. The punch of an arm-strapped shield is likewise much weaker than the punch of a center-gripped shield: when the rim of the shield contacts the opponent, the straps permit it to slide backward somewhat along the arm, robbing the blow of force.

One last note: keep in mind that the distinctions I'm drawing between "offensive" center-gripped shields and "defensive" arm-strapped shields are on a spectrum. Of course a fencer can trap and punch and hook with arm-strapped shields and can use center-gripped shields in a static or purely defensive manner. When it comes to any given fight, a fencer must use any shield however the situation calls for using it (this is one plausible-sounding explanation why some heater and kite shields had so many straps: the different straps provided different ways to grip the same shield, so a fencer could effectively switch between "offensive" and "defensive" mode mid-battle).

medieval warfare, but if any technical treatise was ever penned regarding their use, it has not survived. The historical use of these iconic shields may be inferred through a combination of experimental archaeology and analogy to the shields for which treatises *have* survived. However, as discussed in Chapter 2, this sort of reconstruction is inherently more speculative than working directly from texts. As a result, we shall confine ourselves in this chapter mostly to the shields envisioned by the authors of our surviving treatises.

Shield use is one of those topics in fencing that exhibits clear differences between schools, but the differences are attempts to solve the same basic problem. Shields occupy a large volume of space in front of a fencer. That is their job, after all; it is what makes them such powerful defensive implements. This bulk, however, also presents a problem: a fencer's own shield can get in the way of the accompanying sword. A significant part of skillful sword and shield fencing is learning how to work around the shield in a way so that it is a defense but not a hindrance.

Our surviving treatises offer essentially two solutions to this problem. Escrima comun favors a very active use of the shield. The default shield position curves the shield arm across the body so that the dome of the shield covers the fencer's left side (assuming a right-handed fencer), with the face of the shield facing the opponent. Godinho describes this guard as follows:

> [T]he shield will be in front of the chest, one palm distant, with the arm circled, and the right breast uncovered.
> (Godinho, 2016, p. 55–56)

Note that Godinho specifically calls for the right side of the chest to be unencumbered by the shield. This is not anatomically necessary: as you can easily confirm for yourself, you can hold your left arm far enough across your body that a shield strapped to it would cover your entire chest. However, doing so would significantly encumber your sword arm. Godinho's default shield guard is a compromise between covering the body and leaving the sword free to work.

In escrima comun, however, the shield does not stay in front of the left breast. This is merely a default position. In actual combat, the shield is moved left and right, high and low, to intercept attacks from any side of the body. Godinho again:

> When the opponent throws a *reves* to the head, pass the whole shield to the right side so that it covers all of the face, and you see the enemy on the left side;
>
> Against the *tajo*, the shield is in front of the left side of the face, so that you see the opponent on the right side;
>
> For the *tajo* to the legs, the shield goes above the quillons of the sword, next to the belly, and with a *reves* the same, in a manner that the parries don't have to be taken on the edge of the shield as many have said, but in the body of it, nor do they have to put it in front of the eyes.
> (Godinho, 2016, p. 63)

In other words, escrima comun turns the face of the shield (its "body," in Godinho's parlance) against every incoming attack. A fencer attacked from the upper left curls the shield arm up so that the left fist is pointed toward the sky. This places the head safely behind the face of the shield while still allowing the fencer unobstructed forward vision. When attacked from the upper right, the fencer's shield arm crosses the chest with the fist still pointed to the sky, placing the face of the shield to the fencer's right. Against low attacks, the fencer parries low with the sword while covering the sword hand with the shield. The shield is even occasionally held directly overhead, like a giant steel hat, to protect against overhead blows.

This method of shield use has the advantage of presenting the largest possible area of defensive real estate to an attack. This increases the defending fencer's margin for error. It also maximally constrains an attacker's ability to feint or change directions to circumvent the shield. One of the standard tactics when attacking a shield-armed opponent is to attack high and strike low (or vice versa). For instance, a fencer can attack the left side of the opponent's head in order to get the opponent's shield to move, only to attack lower on the left side of the opponent's body. By meeting every attack with the face of the shield, the defending fencer ensures that an opponent who wishes to change the angle of attack in this manner must circumvent the largest possible distance. This makes the redirection more obvious and forces it to take more time, all of which increases the chances that the defending fencer can successfully react to it.

However, escrima comun's active use of the shield is not without its disadvantages. Consider Godinho's shield parry against a *reves*, or backhand cut, targeting a fencer's upper right quadrant. Godinho prescribes passing the whole shield arm to the right side of the body against such an attack. This forces the sword arm to remain below the shield arm if it is to have any range of motion at all. This in turn limits the counterattacks that the defending fencer can perform: the defending fencer can cut or thrust below the shield arm, but can only attack higher targets by making relatively weak cuts from the wrist or elbow (or by lowering the shield). In a similar manner, if the shield is covering the hand and belly during a low parry, the defending fencer cannot easily make any attacks other than low thrusts. Indeed, thrusts to the legs or the torso are the predominant counterattack in escrima comun when using a shield.

By contrast, Bolognese fencing largely confines the shield to the fencer's upper left quadrant, with the shield arm held strongly toward the opponent. This presents the face of the shield to high cuts against the fencer's left side, just as in escrima comun. In Bolognese, however, the sword is responsible for defending against attacks from other quadrants, and even bears significant responsibility for defending the upper left. Because the shield does not move as actively, and is not held across the chest, the sword arm is free to move almost without hindrance.

The advantage of this method of shield use over that taught in escrima comun is that a shield-armed fencer can make a wider variety of attacks and counterattacks. In particular, backhanded cuts and thrusts are much more common, because the shield arm never constrains the sword arm. But I do not wish to be understood as drawing a simple contrast between "defensive" escrima comun use of the shield and "offensive" Bolognese use of the sword. Indeed, the case can easily be made that escrima comun is actually the more offensively-minded school, as Bolognese makes more frequent use of combined defenses with the sword and shield. Consider the following, Marozzo's seventh play with sword and shield:

> [L]et us propose that your enemy now throws a mandritto to your head or leg, or a riverso or stoccata. In response to each of these, take a big step forward with your right foot to the enemy's left, a bit on the diagonal. As you do so, throw a riverso sgualembrato across his sword arm, and once you've done so, throw a rising falso from beneath your shield into his sword hand.
>
> (Marozzo, 2018, p. 149)

Here, the fencer begins with the left foot and left side forward, pushing the shield toward the opponent. The opponent throws a backhanded cut to the head, which from the shield fencer's perspective comes from the upper right quadrant. Godinho would parry this attack with the shield by passing the shield across the fencer's body to the upper right. Marozzo, by contrast, has the fencer step diagonally to the right, toward the incoming cut, twisting the body from left to right to throw a backhanded cut to the attacker's sword arm. This provides two defenses: the shield fencer cuts the opponent's sword arm, but if that fails, the motion of the fencer's body interposes the rim of the shield against the incoming attack.

It is interesting to remember that this sort of rim parry is just what Godinho prefers *not* to do: after all, since the rim of the shield is circular, the incoming sword may roll down it to strike the defending fencer anyway. This is precisely the sort of danger Godinho prefers to avoid by always taking attacks on the face of the shield. The Bolognese response appears to be that, as in this play, the rim parry is itself a secondary defense. The incoming attack *should* be defeated by the cut to the opponent's sword arm—and, even if the attacker's sword does roll off the rim, the rising back edge cut to the attacker's sword hand will defend against it.

The truth is that fencing systems defy simple categorization as "defensive" or "offensive." This is especially true when fencing with sword and shield. When facing an opponent with a single weapon of similar reach, it often behooves a fencer with sword and shield to aggressively push the attack (though when facing an opponent with a single weapon of superior reach, as we will see at the end of this chapter, even a shield is often inadequate to even the odds). An aggressive approach minimizes the time in which an opponent can work around a fencer's shield to strike the fencer behind it—the topic to which we now turn.

When facing a shield-armed opponent, historical fencing focuses heavily on two targets: the face and the legs. The physical logic behind this is simple: every guard with a shield must leave a fencer's eyes visible (lest the fencer blind himself), and the legs are the hardest part of the body to defend with a shield. The leg target in particular is so important that it can affect the metagame of modern tournaments: I can tell you from experience that shields benefit far more than other weapons when sparring rules disallow the leg as a target (or prevent fencers from being eliminated by hits to the leg).

Sidebar: Why not cut the body?

One fact that often surprises non-fencers is the dearth of cuts to the body (or torso) in historical fencing treatises. While cuts to the body are historical, they are strongly disfavored by most systems, and almost entirely absent when fencing with one-handed swords, regardless of system.

To understand why requires experience test cutting targets with live blades. One-handed sword cuts are quite capable of severing necks and limbs but they lack the momentum and stability required to bite deeply into torsos with regularity (regularity being the key word). When attacks to the torso are made with one-handed weapons, it is almost always with a thrust. One-handed thrusts can reliably penetrate the torso to reach the vitals; one-handed cuts do not.

> This is an excellent example of the importance of an approach to historical European martial arts that unifies sparring experience, test cutting, and close reading of historical sources. It is quite easy to cut the body in one-handed fencing (so easy, in fact, that it can be hard to train fencers to strike the legs instead of the belly or ribs). However, a close read of sources suggests that such body cuts would have been considered poor form, and test cutting helps to explain why.

Attacks against the face or eyes of a shield-armed opponent are principally important as threats. As the shield is typically held quite close to the eye line regardless of fencing system, it is trivially easy for a shield user to bat away a thrust or a cut to the face. But this still requires some movement of the shield, during which it cannot be closing additional attacks and is moving even further away from the legs. During the shield's defensive movement, the attacking fencer can strike the true target. Manciolino begins his discussion of sword and shield fencing with just such a sequence of attacks:

1. Push a thrust with your left foot forward.
2. Gather back with your left foot, extending your sword behind you, then pass forward with your right foot feinting a mandritto to his head. As he lifts his rotella in fear of your cut, you can do one of two things:
    a. Hit his leg with a riverso.
    b. Pass forward with your left foot and push a thrust to his flank; then retreat with a backward jump.

(Manciolino, 2010, p. 136)

Here, Manciolino begins with the simplest of all attacks against a shield-armed opponent: a stab to the face, above the shield. His second attack presents a variation on the theme, attacking the opponent's head with a forehand cut, which causes the opponent to parry upward with the shield. Manciolino recommends making this feinted cut with an exceptionally large circle of the arm, beginning it by "extending your sword behind you" so that the cut seems to be descending almost vertically as the sword swoops back and around. This has two effects: it ensures that the opponent sees the attack coming, and it allows the attacking fencer to change the direction of the feinted cut with a very small motion of the wrist at the apex of the cut, when the sword is directly overhead. Seeing the obvious attack, the defending fencer confidently raises the shield to cover the upper left, clearing the way for the attacker to strike beneath the shield at the opponent's leg.

Godinho's first play of sword and shield likewise emphasizes the importance of low attacks against a shield with a back-and-forth sequence that showcases some of the active shield use so characteristic of escrima comun:

> If the opponent tests the sword, before it arrives, give a *reves* in the leg, not taking the shield from its place. The enemy can lift the leg, so that it passes in vain, and thrust nails-up to the chest, which the opponent can defend by the same edges with a nails-up thrust. The opponent can undo this thrust to his right side with the shield and give a nails-down thrust in the thigh, putting in the left foot, covering with the left face of the shield.
> (Godinho, 2016, p. 56)

In this sequence, both fencers are armed with sword and rotella. The first fencer attempts to "test" the opponent's sword by gently crossing swords, to see how the opponent will react. The opponent gives a backhand cut to the legs before the swords even touch (a good example of how escrima comun, for all its emphasis on shield use, also often exhibits a hair-trigger aggressiveness). The first fencer moves the leg to avoid the cut, responding by thrusting at the right side of the opponent's chest with the palm turned upwards (similar to the hand position referred to as fourth or *quarte* by modern fencers). The opponent fires back with a palm-up thrust to the first fencer's right chest (recall that in escrima comun the default shield guard leaves the right chest uncovered), crossing the line of the incoming thrust and knocking it toward the fencer's right. The first fencer knocks this counter-thrust out of the way in turn by crossing the shield to the right, simultaneously turning the sword hand over so that the palm is facing down and stabbing the opponent in the thigh.

Both these sequences exhibit the importance of leg attacks in historical shield combat. They also demonstrate that a committed attack to the legs is unlikely to succeed as an opening gambit. The main reason for this is reach. Swords are attached to the body through the shoulder, so geometry dictates that a sword will be able to reach targets close in height to the shoulder (such as the face or chest) before it can reach targets far from the shoulder (such as the legs).

One consequence of this is that shield fencers may attempt to target the chest instead of the legs, notwithstanding the fact that the chest is the part of the body best defended by the shield. Sometimes the allure of striking at the longest possible range (and, thus, the first possible moment) is simply too great to ignore.

Both of the plays we just looked at include attacks to the chest. As a practical matter, however, attacks to the chest are often harder to land than attacks to the leg, simply because they necessarily pass closer to the defending fencer's shield (recall the cross-body shield parry against a thrust to the chest in Godinho's sequence). Godinho often compromises by targeting the belly rather than the upper chest. As he puts it, "although it is mentioned that you give the thrusts in the chest, those in the belly or lower are much better, because those that go to the chest will be easy for the opponent to undo with the shield" (Godinho, 2016, p. 61).

No wonder, then, that torso attacks against a shield-armed fencer often require greater subtlety than the simple high–low combination of leg attacks. Either the attacking fencer must be especially mindful of her timing and the defending fencer's motion, or the opponent's own weapons must be securely occupied.

Godinho gives an example of such a thrust in the following sequence:

> [I]f the opponent commits a nails-up thrust to the chest, take the sword on your shield, raising it at the same time to the head, lifting the arm well, and give a nails-up thrust in the belly, not entering more than enough that the thrust reaches. The opponent can defend this thrust by undoing it with the shield to the left side give a nails-down thrust in the belly, putting in the foot and placing the shield to the left face.
>
> (Godinho, 2016, p. 56)

Here, the attacking fencer only successfully lands a thrust to the opponent's belly after taking both of the opponent's weapons out of commission. The attacker occupies the opponent's shield by thrusting at the chest, which the defender defeats by raising the shield over the head like a roof, carrying the attacker's thrust high. The opponent then counterattacks with a thrust to the belly, the attacker sweeps it out to the left with the attacker's own shield. Only then, with the defender's arms separated and both sword and shield occupied, does a second thrust to the defender's belly strike home.

Separating the weapons (or allowing them to be separated) in this manner is a danger against which sword and shield fencers must be constantly on guard. Whenever the sword and shield begin to drift (or are forced) apart, the opponent may attack between them. This can be used as a lure for an unwary opponent, but it is a surprisingly easy mistake to make. Shields are large and powerful defensive arms, and it is easy to forget that they do not make one invincible. It is the mark of a novice shield fencer to feel so secure behind their shield that they neglect the position of their sword, inadvertently leaving themselves open.

So far, we have discussed defensive use of the shield only. This may surprise you, as popular media is increasingly aware that the punch of a shield rim can be a devastating attack. In the specific case of rotella fencing, however, there is a very good reason for this: the straps.

Center-gripped shields are held by a rigid bar, which allows the shield to be used as a very heavy bludgeon. The German longshield dueling tradition seems to have made significant use of this feature, to the point that duelists sometimes discarded their weapons entirely and attacked solely with their shields, using them in two hands.

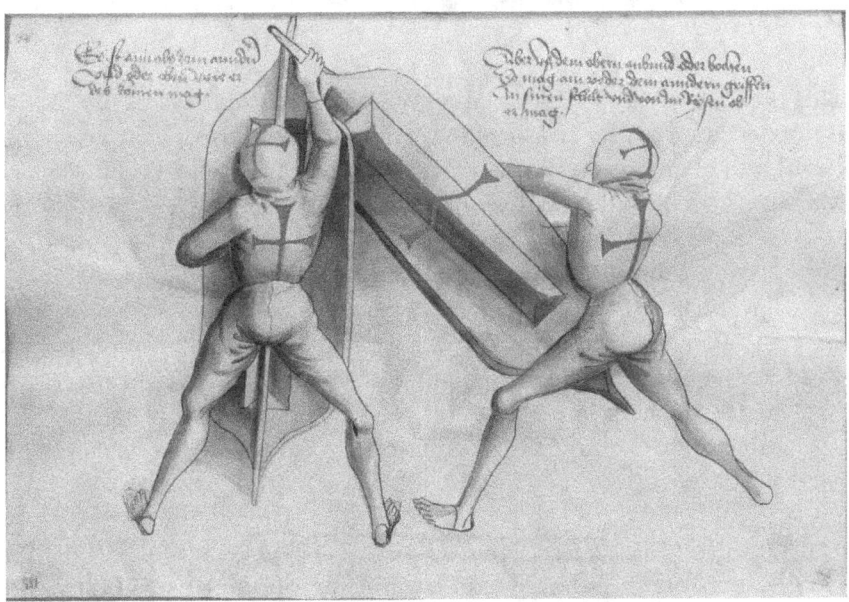

Arm-strapped shields, however, are a different beast. When a fencer punches with an arm-strapped shield, the shield tends to shift backward on his arm due to the non-rigid grip strap. This robs the blow of force. For this reason, it is probably no surprise that neither escrima comun nor Bolognese include a single striking technique with the rotella.

> **Sidebar: Do arm-strapped shields really need straps?**
>
> An arm-strapped shield has at least two straps: one near the center of the shield that goes around the fencer's forearm, and a second near the rim that is used as a hand grip. You may be wondering why, if straps rob a shield punch of force, fencers did not replace the grip strap with a rigid handle. This is certainly possible, and modern fencers whose sparring rules permit full force shield punches (usually against plate armor) often make this modification.
>
> However, replacing the grip strap with a handle means the hand grip is no longer adjustable. An adjustable grip strap can be loosened to allow the shield hand some freedom of motion—enough to pick up or hold objects (including the haft of a spear, as we'll see in Chapter 10). Replacing the grip strap with a handle robs the shield of this utility, which is a major feature of arm-strapped designs.

This is not to say that the rotella cannot be used offensively, only that it is not designed to give heavy blows. For instance, the grip strap is stable enough to push on an opponent's arm, temporarily disabling it. This is Marozzo's second play with sword and shield:

> [T]hrow a thrust to his face as you step forward with your right foot toward his right side. Out of fear of this thrust he'll raise his sword, and all in one tempo, you'll take a big step forward with your left foot toward his right side and put your shield under his right arm, that is, into his sword arm, and as you put your shield there, throw a mandritto to his right leg and make your right foot follow behind your left one.
>
> (Marozzo, 2018, p. 147)

Here the attacking fencer works around the opponent's right side so that the attacker can pivot and thrust the shield into the attacker's sword arm. In Marozzo's play, the defending fencer's sword arm is

raised to push the attacker's thrust overhead, and the attacker's shield is shoved into the defender's armpit to keep that sword arm out of play while the attacker cuts the leg. As a variation, the technique can also be used to shove the sword arm into the opponent's body for a similar effect. While neither of these options has the satisfying brutal impact of a shield punch, they remind us that an attack with the shield should always support an attack by the sword.

Before we leave sword and shield fencing, we should discuss the use of the two other shields for which we have surviving technical material: the *imbraciatura* and the body-length center-gripped dueling shield that modern HEMA fencers call the "longshield."

The imbraciatura is a long, concave, oblong shield strapped to the arm and held by a horizontal handle, featuring a spike at its lower end. Marozzo is the only surviving author to treat this weapon, but it offers several differences from the rotella that may be illuminating.

The most obvious difference between a rotella and an imbraciatura is that the latter is long enough to cover the leg. This presents an immediate problem for the standard sword and shield theory we have looked at so far, which relies heavily upon being able to strike the leg. The next most obvious approach, based on rotella fencing, would be to attempt to attack between the sword and imbraciatura. No wonder, then, that Marozzo advises the imbraciatura fencer that "you must set yourself ... with your sword and your immbraciatura close together" (Marozzo, 2018, p. 197). If the savvy imbraciatura fencer is careful to keep weapons together and the shield removes the legs as a target, what is an opponent to do?

Marozzo's first piece of advice is to let the opponent make the first move:

> [W]hen he throws one of the aforesaid blows, put your imbraciatura into whatever blow he throws, stepping forward to his left side in that tempo with your right leg and throwing a thrust to his face or flank as you step.
>
> (Marozzo, 2018, p. 198)

In other words, when the opponent's shield is too large to allow the usual options of feinting to cause the shield to uncover a target, the savvy fencer must tie up the opponent's weapons by allowing the imbraciatura-armed fencer to attack first.

It is not hard to imagine this advice being rather depressing for one facing an imbraciatura (not to mention a strong argument for acquiring an imbraciatura of one's own!). Marozzo does offer advice on how to press the attack against one armed with an imbraciatura. Interestingly, they both involve getting very close to the opponent. Here is one:

> [F]ind your enemy with a riverso traversato, without stepping. Once you've thrown the riverso, thrust a punta riversa to his right side, stepping forward with your left foot in that tempo. As soon as you've thrown the punta riversa, hit the *penna* of your imbraciatura forcefully into the underside of your enemy's, stepping forward to his left with your right foot as you do so. In the tempo wherein you step, extend a falso impuntato to his face, along with a mandritto to the legs.
>
> (Marozzo, 2018, p. 198)

This is a complicated sequence of attacks (a testament to the difficulty of pressing the attack against such a large shield) in which the fencer ends up shield-to-shield before she can land an attack. The play begins with a backhanded cut and backhanded thrust toward the opponent's sword side, harrying the opponent so that the attacking fencer can close the distance. The real attack begins when the attacker hits the lower part of the opponent's shield with the *penna* (whether this means the lower rim or the face of the shield is uncertain; the Italian is ambiguous) of the attacker's own shield, thus levering the opponent's out of position. This exposes the opponent both to a falso impuntato (a thrust delivered like a right hook) to the face and a cut to the legs.

Marozzo's second attack similarly involves forcefully controlling the opponent's shield:

> [C]ast your right hand into your enemy's imbraciatura, and using either your pommel or your hand, grab it from above and pull it strongly to you so that you'll be able to knock him over with little effort by driving the bottom part of your imbraciatura into his left shin so that he's unable to stay on his feet. But bear in mind that if he backs up, you'll be unable to do this presa, in which case throw a riverso tondo to his face, followed by a mandritto traversato.
> (Marozzo, 2018, p. 198)

Here, the attacking fencer controls the opponent's shield even more directly, by seizing its top rim and pulling it forward. The attacker can pin the opponent's leg with the spike of the attacker's own imbraciatura, thus toppling the opponent forward (the imbraciatura, recall, is strapped securely to the opponent's arm) or simply give backhand and forehand cuts to the face once the shield has been pulled out of position.

The general trend of imbraciatura fencing, then, is that fencers must get very close to each other when their shields become very large. This is a trend that we will see continued as we discuss our third and final type of shield, the longshield.

The imbraciatura was a general weapon for dueling as well as warfare; Marozzo discusses its use both one-on-one and against polearms. The longshield, by contrast, appears to have been exclusively a dueling weapon. The weapon itself is a body-length shield held by a long pole that ran the length of the shield. This length gives it its modern name (the German word used in treatises discussing it is simply *schilt*, or shield). All illustrated longshields feature spikes at the top and bottom, and many feature a variety of additional spikes or hooks at the corners. I am aware of no surviving antiques (like most shields, longshields were made of wood), but modern reconstructions end up quite heavy, in the neighborhood of 10 pounds or more.

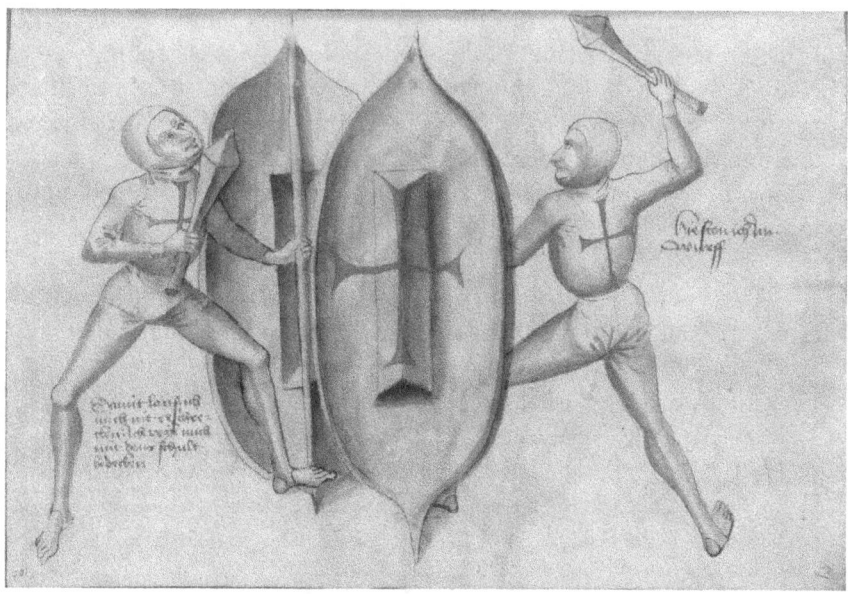

This bizarre weapon is treated only by German fencing treatises and seems to have been strictly used for an especially ritualistic form of duel. Medieval treatises always illustrate it being used in a formal judicial dueling yard, and the combatants are shown wearing ritual white outfits with a red cross upon them that Talhoffer says double as burial shrouds. The longshield is occasionally shown accompanied by a one-handed sword, but more often by a small mace or club with a diamond-shaped head. This mace's form is remarkably consistent across illustrations, suggesting it too may have been part of the longshield's ritual tradition. The use of the mace likewise suggests a very formal environment. Paulus Kal includes a technique in which a fencer throws his mace in a longshield duel "and call[s] [his] marshall to give him another" (Kal, 2019). He introduces another technique with the words "if you have used all your maces" (Kal, 2019). Advice such as this makes sense only in a strictly regulated environment. Yet despite the highly ritualistic nature of the weapon, longshields do have some things to teach us about shield usage in general.

The great size of a longshield means that it presents much the same challenge as an imbraciatura: it can defend virtually the entire body. As with the imbraciatura, the advice of longshield treatises is universally that a proactive fencer must close nearly to grappling range to overcome this defense.

The center-gripped nature of the longshield, however, offers different methods of manipulating it than the hybrid arm-strap-and-grip nature of an imbraciatura. As we have seen, a fencer primarily manipulates his opponent's imbraciatura by knocking it back or pulling it forward. A longshield, however, is not strapped to the arm, but held by a rigid vertical pole. Moreover, a longshield's bottom spike is often planted in the ground, allowing the wielder to use it as a portable wall. This means that a longshield is most easily manipulated by rotating it around its vertical axis, causing it to pivot like a door.

A number of offensive longshield techniques rely on this pivot action. The shield can be moved across the body and pivoted so that its face is flat against an opponent's weapon, trapping it while the attacking fencer strikes behind. We saw a similar action from Bolognese rotella fencing, in which the rim of the shield is pushed into the opponent's sword arm. A longshield, however, can do so with the face of the shield, effectively trapping the opponent behind a wall from which the attacking fencer will emerge behind.

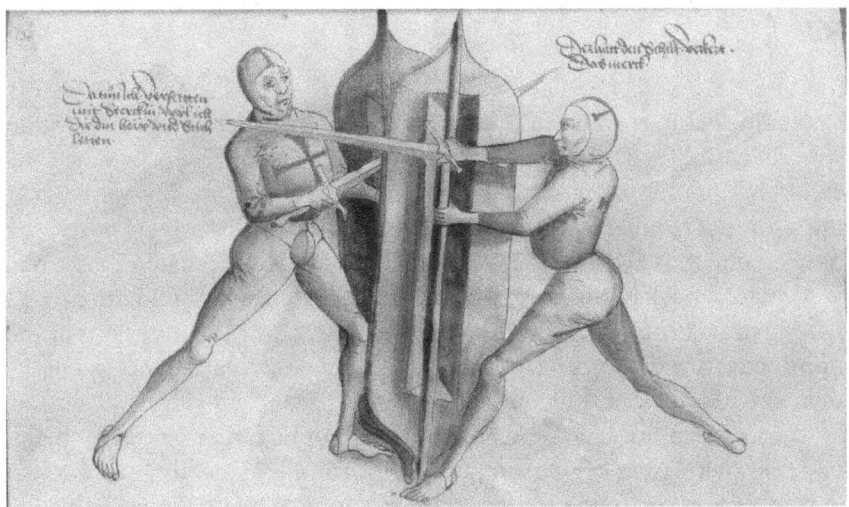

Another method of attacking a longshield is to kick the face of the shield so that it turns inward in the wielder's hand. This exposes the opponent's left side to attack, which has the double advantage of allowing the fencer to put as much distance as possible between himself and his opponent's weapon and attacking what is ordinarily the opponent's best-defended side. To protect the kicking leg from being struck, the attacker's own shield is rotated to face left, so that the kicking foot ends up safely between two shields.

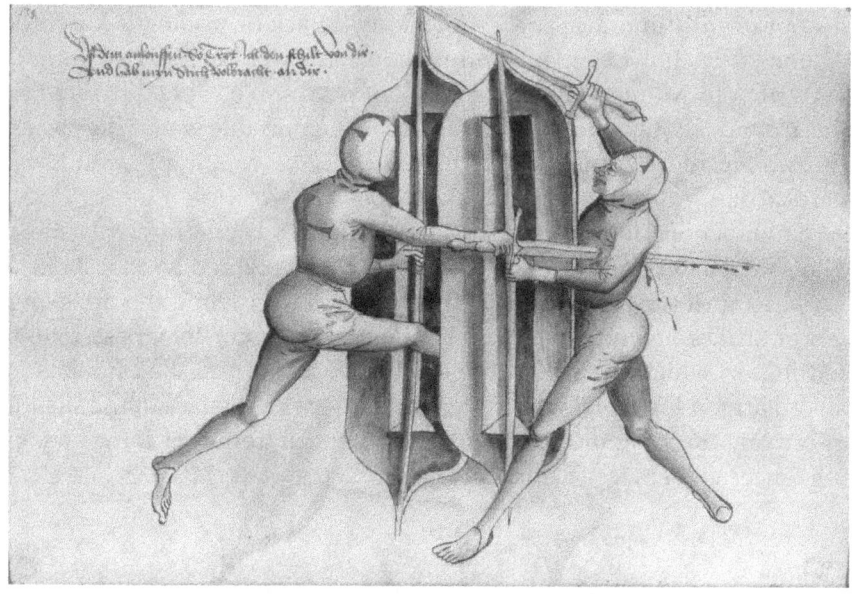

Shield-kicking is a fun, attention-grabbing technique that often sounds easier to execute than it is. The technique only works when the shield is held vertically, so that it will rotate around its center grip; kicking a longshield that is slightly off-axis does not necessarily produce the desired effect. It also requires the attacker to be able to deliver a strong, well-placed kick while moving the attacker's own shield to protect the leg and still follow up with a thrust or cut quickly enough that the opponent cannot simply back out of distance or rotate the opponent's own shield back to protect the left side.

These techniques work well with any large center-gripped shield, whether oblong like a longshield or round, and they neatly illustrate the importance of the rotation that a center-gripped shield allows. While an arm-strapped shield like a rotella has a fixed orientation relative to the shield arm, a center-gripped shield can be presented with the face forward, left, or right. This variable orientation can be used offensively, such as by trapping an opponent's weapon. It can be used defensively, not only to parry weapon strikes but to create a channel for attack, as when it protects the kicking foot. It can also be used against the opponent, as when the shield is forcibly rotated inward to expose the opponent's shield side.

There is one last feature of longshields that is unique to this form of weapon, which is its ability to be wielded in two hands like a polearm, hooking and stabbing with the shield's various projections. In fact, a large number of extant longshield techniques are performed with the shield held in two hands and no other weapons at all.

Fencing with the longshield two-handed largely eliminates the opportunities to rotate or otherwise manipulate it around the vertical axis, because the shield is so rarely held vertically. However, the weapon's ability to rotate in the hand can still be used to good effect both offensively and defensively.

A technique known as the heart thrust illustrates both. In Paulus Kal's version of the heart thrust, the defending fencer holds the shield diagonally across the body. The attacking fencer charges forward, shield held vertically to the fore like a wall. The attacker then reciprocates the diagonal angle of the opponent's shield so that the two shields form

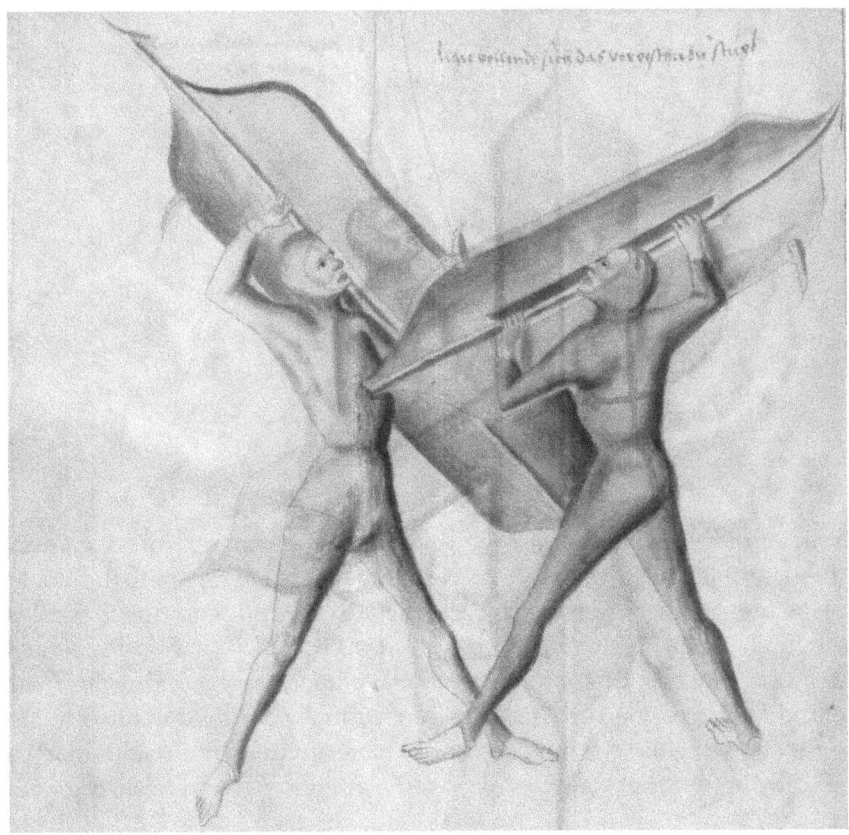

an X and scoops the attacker's shield up over the head, flipping the whole weapon around in the opponent's hands. While the opponent has momentarily lost control of the shield due to the shifting weight, the attacker stabs down to the opponent's heart.

Talhoffer illustrates a different way to achieve the same result. In Talhoffer's version, the defending fencer holds the shield two-handed low and on the left side, deliberately offering the upper right quadrant as a tempting target to the opponent. When the opponent's shield stabs down, the defending fencer raises the shield overhead and flips it over so that the face of the shield defends the right side. The opponent's thrust thus slides harmlessly by, while the defender drives the shield's spike down into the opponent's heart.

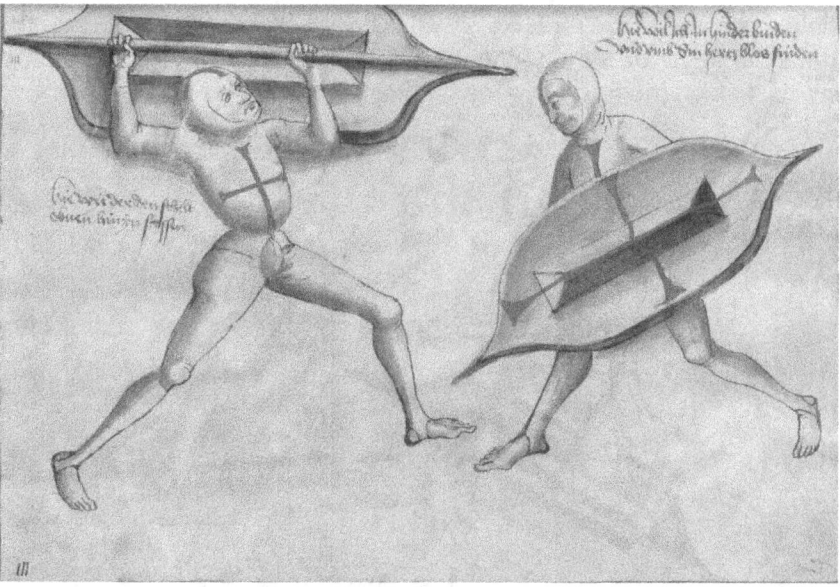

There is more research to be done about the context of longshield duels, so it is difficult to say with certainty where fencing with the shield alone fits into the larger tactical scheme. Kal's cryptic comments such as "if you have used all your maces" suggests a known set of rules, which likewise suggests that a fencer may have had reasons under the rules of the duel to expend whatever hand weapons permitted in the dueling ring—reasons that did not necessarily have anything to do with martial efficacy in the strict sense. However, the attention given to fencing with

QUEEN OF SWORDS

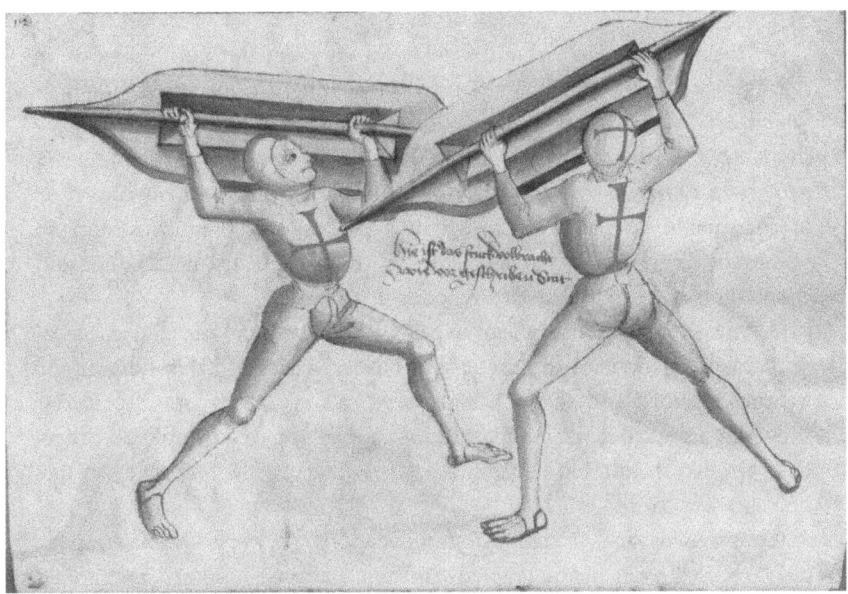

the shield alone in longshield sources is an evocative reminder of how effective any shield can be as a weapon. Of all the companion weapons for swords that we will discuss in this book, there can be little doubt that the shield is the most powerful. It is, as Godinho says, "queen."

Which would win: sword and shield or longsword?

I get a lot of questions about whether sword and shield or longsword is the superior weapon combination. I am not entirely sure why. My best guess is that this question stems from the suspicion that as European armor progressed, longswords eventually supplanted the sword and shield as the knightly sword combination *par excellence*. There is thus, perhaps, a bit of time machine curiosity inherent in the question: was the new way really better than the old?

Or maybe the question is less academic than that. I certainly spent many a night (and many a lunch period) debating whether the extra damage of a two-handed sword was really worth giving up the extra points of armor class that a shield provided.

I am aware of no historical sources that discuss this question directly. However, sparring experience suggests that the answer is fairly clear.

Without armor, sword and shield is a very difficult combination for a longsword to defeat. It is essentially cheating: an unarmored fight in which one of the combatants does, in fact, have armor (the shield). To the extent that the longsword supplanted the sword and shield as the preferred sidearms of elite warriors due to the progression of armor (a reasonable conjecture, though not one on which fencing masters comment directly), it is hardly fair to pit the two against each other and then take away the longsword fencer's armor.

A fencer armed with a longsword is, of course, perfectly capable of implementing most of the tactics we have discussed in this chapter for fighting a sword and shield. Longswords can feint high and attack low; they can attack between separated weapons; they can hook or otherwise grapple shields. However, there is a significant difficulty that may not be readily apparent.

Longswords are, as two-handed weapons go, very short. A spear, poleaxe, and even a greatsword can fence against a sword and shield with advantage because their length allows them to change targets from a safe distance (indeed, as we will discuss in Chapter 8, Godinho explicitly recommends against engaging a greatsword with a sword and shield). As mentioned, a longsword can feint high and attack low too, but only from a distance at which the shield fencer can also attack—and, of course, the longsword fencer has no second weapon with which to defend while doing the relatively laborious work of deceiving the opponent's shield out of position.

A longsword's main advantage against a one-handed sword is its superior leverage. A shield helps to obviate this weakness on the part of the one-handed sword, because a longsword does not have nearly enough leverage to overcome a shield. Thus, so long as the shield fencer can oppose the opponent's attacks with the shield or otherwise avoid remaining at crossed swords, the longsword is robbed of much of its strength.

Of course, longswords remain very fast, as we discussed in Chapter 3—far faster, in practical terms, than a one-handed sword. This is an important factor for the shield fencer to take into consideration. It is easy for a shield fencer who is unused to facing longswords to mis-time actions. The first time I ever fenced with a longsword against a sword and shield, this was exactly my experience. My opponent was unused to how quickly longswords can change direction, and I found it very easy to strike around his shield no matter how aggressively or cautiously he fenced. If I had written this book immediately after that bout, I would be giving a very different answer.

Subsequent experiences, on both ends of this matchup, have been less naive about a longsword's capabilities. The shield fencer cannot afford to be lazy against a longsword. But the matchup is not otherwise in the longsword's favor.

When armor is present, the tables turn. As we will discuss in Chapter 12, using a sword against armor is an involved process that often results in grappling. Effective one-handed sword use against armor is quite difficult; all of our extant sword techniques against armor assume the use of two hands to maneuver the blade and stabilize it for the hard thrusts necessary to threaten mail. Thus, a longsword fencer in full armor would need to fear the opponent's sword far less than he otherwise would, and would find it much easier to close to a range at which the opponent's shield could be wrenched or muscled aside.

You may be wondering about the effect of shield punches on armor. A full treatment of blunt impact on armor can be found in Chapter 12. For now, we can make two observations. The first is that, as we have discussed in this chapter, not all shields can punch effectively. The second is that, at least with respect to plate armor, the effectiveness of blunt force trauma compared to stabs is subject to frequent exaggeration and misunderstanding. A strong punch from a center-gripped shield is not an attack that a plate-armored fencer can simply ignore, by any means—indeed, it is possible (albeit unlikely) for a strong shield punch to knock an armored opponent down completely. However, it is not likely that the ability to punch with a shield would be sufficient to overcome the other advantages of a longsword in an armored contest.

CHAPTER SIX

The forgotten: sword and cloak, sword and dagger, and two swords

> *I remember hearing it said that the only arms were sword alone, and sword and shield, and there are not many times that there was notice of sword and dagger.*
> —Domingo Luis Godinho (Godinho, 2016, p. 71)

In the preceding two chapters we have dealt with one-handed swords accompanied by various types of shields, from the ubiquitous buckler to the exotic German longshield. In this chapter, we discuss more unusual companion weapons: the cloak, the dagger, and the second sword.

I should start this chapter with a confession: there is a possibility that the weapon combinations we're going to discuss in this chapter are not, strictly speaking, medieval. Fencers normally associate these weapons with the Renaissance, even when they are treated by systems that have medieval roots. Certainly, the bulk of our extant material on fencing with a sword accompanied by a cloak, dagger, or second sword comes from Renaissance authors.

Why this might be is something of a mystery. Cloaks, daggers, and two swords were all physically present in the Middle Ages, after all. It may have something to do with the social respectability of different companion weapons. For instance, dall'Agocchie writes of different companion weapons that "the unaccompanied sword is accepted everywhere, and is in greater use, and can always be had more easily, but this is not so with other ones," while Manciolino, writing 41 years earlier, notes that "while you do not have your *rotella* or buckler with you at all times, your sword can always be at your side" (dall'Agocchie, 2018, p. 14; Manciolino, 2010, p. 76). These references suggest that different companion weapons might have been more or less acceptable to carry at different times or in different places (it is also interesting to note that the two companion weapons Manciolino refers to are both shields). It is certainly not hard to imagine that carrying a rotella about town might have drawn unwanted attention. As we discussed in Chapter 4, it seems that bucklers were considered more acceptable in the Late Middle Ages than in the early Renaissance, leading to a surge in popularity of other, non-shield companion weapons.

However, there is not a complete dearth of late medieval evidence for the non-shield weapon combinations. We do have a few scattered references from escrima comun to cloaks, daggers, and two swords as companion weapons in the 15th century—at least, according to Pacheco, who references these weapon combinations in disparaging quotes of the now-lost treatises of both Pedro de la Torre and Jaime Pons de Perpiñan, both of which texts dated to circa 1474.

Regardless, the non-shield companion weapons provide a useful perspective on the pros and cons of the one-handed fencing we have looked at hitherto. They are also fascinating in their own right (who among us has not wondered about the practicality of wielding two swords at once?), and discussing them will allow us to address some misconceptions that we have not yet had occasion to put to bed. And so, leaving these chronological quibbles behind, we'll begin by discussing the combination of sword and cloak.

Somebody asked me once at a public demonstration whether historical swordsmen really wore cloaks as often as they do in fantasy art. The question betrayed the common misconception that there is something about the cloak that particularly lends itself to expert swordplay. This is not true. An expert swordsman can, of course, make good use of any companion weapon. That is part and parcel of being an expert. The real utility of a cloak, however, is that it was everyday outerwear. To continue Manciolino's thought, a fencer might not always have his buckler or rotella, but he would almost certainly have a cloak simply by virtue of being dressed. Sword and cloak is, in other words, the historical equivalent of wrapping a jacket around one's arm as an impromptu shield (strengthening this analogy is the fact that authors who treat the sword and cloak discuss both how to remove it quickly, as one would have to do in a spontaneous encounter, and how best to take hold of it with more care, as one would do at the start of a duel).

Sword and cloak fencing is treated by both Bolognese and escrima comun. There are two primary ways to use a cloak recognized by both traditions (though, as we will see, not all masters approved of them equally): as an improvised shield and as a distraction. Each of these techniques had, in turn, two main subdivisions.

The majority of extant cloak techniques involve parrying with the cloak. The simplest way to do this is to parry an opponent's cut directly, essentially using the cloak as if it were a buckler. To do this, the cloak must be fully wrapped around the arm so that the arm is protected by as many layers of cloth as possible (when fencing with a rapier and cloak, a fencer can allow the cloak to "fall" off the arm to create a curtain that offers some protection to the lower body; against heavier cutting swords, every inch of fabric is needed to protect the arm itself). There are multiple ways to accomplish this; here is one example from dall'Agocchie:

> So, when you have your cape about you, let it fall down from your right shoulder as far as the middle of your left arm, and then turn your left hand outwards, collecting the cape upon your arm.
> (dall'Agocchie, 2018, p. 60)

The fencer shrugs out of the cloak, then makes large windmilling circles with the left arm to wrap the cloak around it. With a bit of practice, this creates an obscuring (and eye-catching) fan of fabric and ends up protecting the cloak arm from fist to elbow. Once the cloak is wrapped around the arm (or "embraced," as the Bolognese put it), it can be used much like a buckler, or textile armor, to block a cut. As dall'Agocchie continues:

> If … he throws a mandritto to your head, step forward with your left foot and defend yourself from that with your cape, thrusting a punta riversa to his chest in that instant.
> (dall'Agocchie, 2018, pp. 60–61)

In this play, the fencer begins with the right foot forward, with the sword held low on the right side and the cloaked arm extended toward the opponent (you may recognize in this stance the basics of Bolognese sword and buckler work). As the opponent throws a descending cut at the fencer's upper left, the fencer steps forward and to the left, blocking the cut with the cloak while simultaneously stabbing through the opponent's chest.

The obvious danger of treating the cloak like a buckler (or dagger), of course, is that most companion arms are made of wood or steel … and cloaks are not. Just like modern jackets, cloaks were made of a variety of fabric weights and types, and a variety of lengths. A short, lightweight cloak intended primarily for fashion did not provide the same cut protection as a long, heavyweight cloak intended to protect from inclement weather. Relying on a cloak to stop a sword cut was thus a risky proposition. Godinho recommends against it entirely, adding a grisly detail from his own experience:

> [I]t is not permissible to crudely take the opponent's blows on the cloak, because although sometimes it doesn't do damage, other

times it does notable damage; I have seen arms mangled from having taken blows on the cloak.

(Godinho, 2016, pp. 113–114)

Even dall'Agocchie adds a strong cautionary note:

> [T]here's a difference when it comes to parrying, because the cape can be cut and pierced, whereas the dagger cannot. Therefore I want to advise you that when you parry either mandritti or riversi with your cape in defense of your upper body, you must parry your enemy's sword below its midpoint, and before the blow has gained force.
>
> (dall'Agocchie, 2018, p. 60)

In other words, a cleaving cut *could* shear straight through a cloak even if it was wrapped around the arm multiple times (I can confirm from my own cutting experience that a medieval sword is eminently capable of shearing through multiple layers of wool). Indeed, one of the major paradigm shifts between sword and buckler fencing and sword and cloak fencing is understanding that the cloak will not reliably protect against a cut or thrust aimed directly at the cloak arm. This means that the off hand and arm are targets when protected by a cloak in a way that they are not when protected by a buckler.

Sidebar: How sharp were medieval swords?

One of the more annoying myths I encounter about medieval swords is that they were not very sharp, or even that they were deliberately left dull and meant only for battering opponents in armor. This is a myth with a long pedigree, so we should set the record straight: medieval swords were *aggressively* sharp.

So far as I can tell, the idea that European swords were blunt stems from the 19th century, as a result of a few factors. For one thing, European military sabers of the time often *were* blunt. Partially this is a result of Europe being in the process of losing its "sword culture," as a result of which not everybody who owned swords had any idea how to properly sharpen them (a state of affairs that persists today). For instance, although the British adopted several official styles of saber use throughout the imperial period, they never adopted an official method of sword *sharpening*. As a result, even when the order came down to sharpen sabers, the actual quality of the sharpening was highly variable. Sometimes individual servicemen had to sharpen their own weapons, with no formal training in how to do so. Sometimes a unit's weapons were sent to an outside source for sharpening, who also might or might not know what they were doing (e.g., a knife cutler, who might or might not know how to sharpen a sword no matter how good he was at sharpening knives; or an armorer). And of course, 19th-century military-issue swords were often stored and transported blunt, for the same reason that guns are stored unloaded. You can easily imagine the logistical (and medical!) difficulties of moving hundreds of sharp blades in the custody largely of young men. Consequently, if you buy an antique 19th-century sword today, the odds are good that it will be blunt. Then again, if you buy an antique 19th-century musket, the odds are good that it will be unloaded. Neither should surprise you.

But we shouldn't overstate the incompetence of Victorian sharpening. Plenty of Victorian swords were quite sharp (and are still quite sharp, if you happen to find an antique that was sharpened in anticipation of seeing action and left in that state), and Victorian swordsmen had a tradition of performing cutting feats that required razor-sharp swords. Another factor in the stereotype is the fact that European armies (especially the British) tended to issue sword-bearing troops with metal scabbards—which, while far more durable than wood-and-leather scabbards (and thus cheaper in the long run for whoever was bankrolling the regiment), also quickly blunted the edge of a sword even if it was properly sharpened.

Combine all this with the Victorian sense that modern people do everything better than ancient people, and the result is a lot of people (not all of them swordsmen) assuming that if their modern, scientific society produced blunt swords, medieval swords must have been positively dull! Even in the 19th century there were people who knew better, but so far as I can tell, this is how the myth got started.

Prior to the 19th century (and again, we must remember that even in the 19th century there were properly sharpened and cared-for swords that performed perfectly gruesome feats), European swords were quite sharp. In fact, some medieval swords that have been preserved in the right sort of river mud retain their original edges.

And then, of course, it's perfectly clear that period fencing masters expected swords (i) to be very sharp and (ii) to be used against people in street clothes, against whom a sharp edge would be a significant asset. Test cutting demonstrates that period clothing protects quite well against swords that are not at least sharp enough to shave with (I mean that literally; a properly sharpened medieval sword should be able to shave the hair off your arm). We have already seen some of the more to-the-point examples of what a medieval European sword should be able to do in one swing from Talhoffer, in Chapter 3:

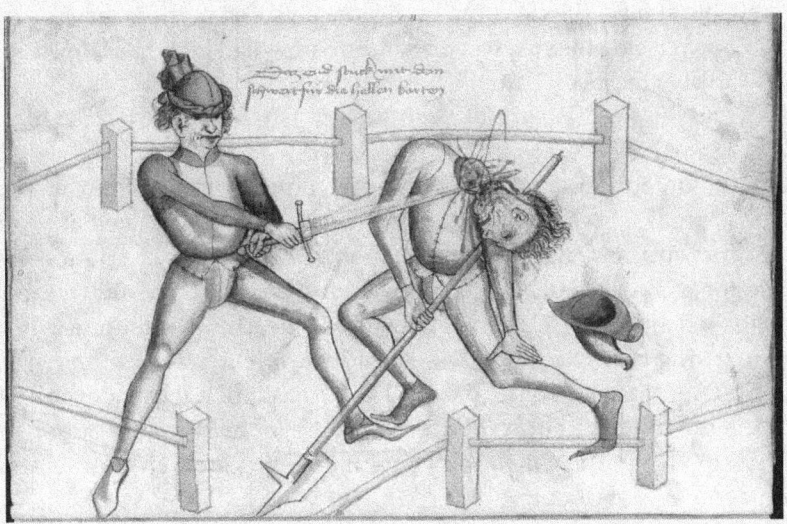

A medieval European sword was expected to be really, really sharp.

As dall'Agocchie warns, safe parrying with the cloak is a difficult matter of interrupting the cut before it has fully developed, meeting the incoming sword at the base of the blade where the sword is moving slower and thus carries less kinetic energy (and as Godinho warns, failing to properly execute a cloak parry may have life-altering consequences). This type of parry cannot be executed without strongly closing distance, which is another way in which parrying with a cloak differs from parrying with a buckler (in the Bolognese play we looked at above, recall that the fencer stepped forward and toward the opponent's sword side in conjunction with the cloak parry).

In practice, parrying with a cloak while meeting these criteria is significantly harder than simply interposing a buckler against a cut. If done well, however, a direct cloak parry can be a nasty surprise for an opponent. Most sword and cloak fencing is done at a distance, so opponents rarely expect a fencer to fly into short-range to perform this type of technique.

Far safer is the second type of cloak parry, in which an attack is first parried by the fencer's sword and then suppressed by the cloaked arm. This is roughly equivalent to a buckler push, a technique we looked at in Chapter 4. Godinho gives an example of this kind of parry:

> It is advised that the *tajo*, when not being defended with a nails-up thrust at the eyes, is taken on the sword, where it breaks the [blow's] fury. The cloak can be given with speed in the manner of *brazal*, freeing the sword at the same time, and giving a nails-down thrust in the lower parts, or a *reves* in the legs.
>
> (Godinho, 2016, p. 114)

Here, the defending fencer parries an incoming forehand cut with the sword, "break[ing] the [blow's] fury." Only then does the cloaked arm shoot forward and up, "in the manner of *brazal*," the escrima comun name for a direct, dall'Agocchie-style cloak parry. Critically, although the cloaked arm is raised above the fencer's head *as if* executing a *brazal*, the incoming blow has already been stopped by the defending fencer's sword. Thus, the cloaked arm is pushing on a stationary blade, not actually intercepting the cut. With the cloaked arm holding the opponent's sword up and to the side, the defending fencer can then counterattack with a low thrust or cut.

Given the discussion we just finished about swords cutting through cloaks, pressing on a sharp blade with the cloaked arm may seem extremely dangerous. The key difference between this type of parry and the direct parry we have already seen is that a cloak—even a lightweight cloak—has much greater resistance to slicing than to shearing. This is often counterintuitive to new fencers, but it is easily verified through test cutting. In order to slice with a sword (or any other blade), the fencer must be pressing the blade into the target as it is drawn or pushed along the target. Exerting pressure into the target while drawing or pushing the sword is ergonomically difficult with a straight-bladed sword. As a result, it is highly unlikely that an opponent will be able to injure a fencer's cloaked arm simply by trying to saw through the cloak as the fencer is pushing on the opponent's sword with the cloak. The garment itself may be damaged, but the arm is well defended.

The second main method of historical cloak use is to fling (or "cast," as it is often phrased) the cloak at the opponent, either whipping it forward while holding onto one corner, or throwing it like a net. This requires the cloak to be held in a particular way, as it cannot be cast if it is too tightly wound around the arm. Marozzo suggests the following (he is speaking to the reader as if he is a dueling coach preparing a client for a duel in which the chosen weapons are sword and cloak, but the same method can be used by an unassisted fencer):

> At the beginning ... you will turn back his cape by giving a twist to the corner that hangs down on the left side when it's worn as it typically is, borne upon the left arm, as I've told you on other occasions. Once you've turned back that corner of his cape, take it and wrap it around his aforesaid left arm. Then, with your right hand, take the other part of his cape, which is on his right side, turn it over his head, and make him grip it, also with his left hand, turning it, and giving the entire cape a turn around his left arm.
>
> (Marozzo, 2018, p. 127)

The key point here is that the left hand grips both the left and right corner of the cloak, even as the remainder of the cloak is wrapped around the arm. Using this method, the fencer can unwind the cloak from his arm and still have the corners gripped in order to fling it.

Casting the cloak does require that it be unwound from the arm, which means that a fencer must choose between embracing his cloak in "parrying mode" or "casting mode." Of these two, parrying techniques form the majority of extant sword and cloak material. However, casting is the one thing that a cloak can do that no other companion weapon can. Hence, it is understandable that a fencer who casts the cloak might wish to retain a grip on it, so that it can be recovered and cast again, or just rewound about the arm to use as a parrying device once more. Marozzo does just that in the third part of his sword and cloak material:

> [T]hrow a thrust to the enemy's face, advancing four fingers forward with your right foot. This thrust is to be to his inside, so that out of fear thereof he will beat it inwards, toward his left, with his true edge, thereby uncovering his right side. You will then advance toward his right with your left foot, throwing your cape in his face without letting go of the part wrapped around your left arm. Only throw half of the cape, and as you do, throw a thrust to his chest. For your safety, withdraw your left foot behind your right one, and as you take that step, pull your cape back and retreat three or four steps, simultaneously wrapping your cape about your arm.
> (Marozzo, 2018, p. 128)

In this play, the fencer feints with a thrust to the opponent's left, which elicits a parry across the opponent's body. This allows the attacking fencer to circle to the opponent's right, whipping the cloak across the face (if the opponent's sword was still on the opponent's right side, the sword might stop the cloak and the cast might fail). The attacker throws a second thrust while the opponent is blinded, then retreats and recovers the cloak into parrying mode.

While this method of casting permits the cloak to be reused, it can cause the cast to fail in a number of ways that emerge in sparring. When the cloak is still gripped in the fist, a cast has a limited range, so the cast may fall short either by not reaching the opponent at all or by only brushing the opponent and sliding off, rather than enveloping the opponent's head. It is also possible for the cloak to twist itself into a "rat tail" that merely snaps into the opponent's body or head rather than fanning out to obscure their vision. A snap to the head may be briefly distracting, but enveloping the head ("encloaking," in escrima comun's terminology) physically cuts off the opponent's vision and provides a much more reliable window in which a fencer can act.

Throwing the cloak without trying to hold onto it can ameliorate these risks, at the obvious expense of losing the cloak for the rest of the fight. Marozzo gives an example of such a cast:

> [T]hrow some rising falsi into your enemy's sword hand, which you will do to distract him by making him pay attention to those falsi. Then, when you see the tempo, step outwards with your left foot, toward his right side, and throw your cape in his face so that you'll be able to hit him however you please.... This toss of the cape is different from the previous one in that with the first one you didn't let go of your cape, but with this one, you do.
>
> (Marozzo, 2018, p. 129)

The basics of this play are similar to the previous cloak cast. Once again, the fencer distracts the opponent and then works around the opponent's right side to cast the cloak onto the head. This technique is performed at a slightly further distance, as indicated by the fact that it begins with harassing cuts thrown to the opponent's sword hand; the attacking fencer overcomes the greater distance the cloak must travel by releasing it when thrown. If the cloak properly envelops the opponent's face, the attacking fencer has a longer than usual window of opportunity to close and strike. In a sense, this type of cloak cast is the opposite of a direct cloak parry. Where the latter can surprise an opponent with a sudden rush to close distance, a thrown cloak can surprise an opponent from outside of normal sword range.

This capacity for surprise is the cloak's greatest strength as a weapon, but it is attended by serious risks. The most surprising cloak parries are difficult to execute and, as Godinho notes, can fail catastrophically. The most surprising cloak casts can easily miss, depriving the fencer of his companion weapon. The cloak should thus be seen primarily as a companion weapon of last resort, superior to an empty hand but otherwise recommended only by its ubiquity.

The dagger is another matter. As dall'Agocchie noted above, the dagger has a notable advantage over the cloak in that it cannot be cut or pierced. It can also offend the opponent directly, whereas a cloak can only support the attacks of the sword. These two characteristics make it no surprise that the dagger, and not the cloak, became the preferred companion weapon of the rapier during the Renaissance and beyond.

The dagger used in sword and dagger fencing can have many forms. Neither escrima comun nor Bolognese specifies a particular hilt or blade type, nor size (though dall'Agocchie recommends erring on the side of larger daggers to make parrying easier). Marozzo and Manciolino illustrate several different types, from the broad-bladed simple-hilted dagger on Manciolino's cover (similar to an Italian *cinquedea*) to the relatively complex dagger with forward-curved crossguard and a projecting ring to protect the back of the dagger hand in one of Marozzo's illustrations. The precise form of hilt and blade is not important for the techniques that follow, so long as the dagger has a crossguard to assist in parrying. In extremis, a rondel dagger such as we will discuss in Chapter 7 can be used to perform some sword and dagger techniques, but many dagger parries become dangerous without a crossguard.

Both Bolognese and escrima comun contain considerable surviving material on the sword and dagger. In both cases, the fundamentals of dagger usage are the same as the fundamentals of shield usage; indeed, dall'Agocchie notes that a fencer who has mastered both cloak and dagger as companion weapons can "more easily achieve understanding of all the other ones"—so easily, in fact, that he omits any explicit instruction on other companion weapons at all (dall'Agocchie, 2018, p. 41).

Hence, in escrima comun, the dagger moves quite freely to all four quarters of the body to intercept or deflect attacks, just as the buckler and shield do. Bolognese tend to use the dagger more actively than either buckler or shield. Thus, a Bolognese fencer who has mastered the sword and dagger will tend to find that a buckler or shield is simpler to handle.

Neither system utilizes the open stance, sometimes seen in popular art, in which the sword is held on the fencer's right side and the dagger on his left. This stance can feel secure, because the fencer has a weapon on each side. In practice, the fencer is vulnerable straight down the center in the same way as a sword and shield fencer who separates the weapons (recall our discussion in Chapter 5). In order to defend the center with the dagger, the fencer must twist right to bring the dagger across the center, and in so doing twists the sword shoulder away from the opponent. To defend with the sword, the dagger must likewise be twisted away. Neither is ideal.

The Bolognese solution to this problem is to follow the general rule of keeping the sword and dagger on the same side of the body. This creates an incentive for the opponent to attack the undefended side, which, in turn, makes the opponent more predictable. It also primes the fencer to use the hips to power dagger parries. Manciolino illustrates this principle in the following play:

> Set yourself with your left foot forward …
>
> If he attacks you with a mandritto to your head, parry by getting with your dagger into Guardia di Testa; quickly pass with your right foot towards his left side while delivering a mandritto to the opponent's leg or a thrust to his flank.
>
> (Manciolino, 2010, p. 135)

The fencer begins with the left foot forward, hips and shoulders twisted somewhat to the right, sword held at middle height on the right side and dagger held low and across the body, close to the right knee. In other words, both weapons are held to the right of the fencer's body and at slightly different heights, with the sword somewhat higher and more to the right than the dagger. The opponent attacks with a descending cut from the right side, toward the defender's left side. The fencer defends by stepping to the right with the right leg, so that the right foot ends in front. At this point, three things happen at once. In the course of taking the step, the fencer's hips turn from right to left. The dagger shoots up and to the left, catching the incoming attack between its blade and crossguard. The fencer's sword, meanwhile, attacks with a forehand cut or thrust from right to left. The rightward step moves the fencer away from the incoming attack and lends the power of the hips to both the dagger and the sword as they traverse the body to either stop the opponent's attack or strike the opponent, almost like turning a wheel with both hands. This turn of the wheel to power both parry and attack is the essence of Bolognese sword and dagger fencing.

Escrima comun, as should be no surprise by now, moves the dagger in whichever manner can intercept an attack. Godinho is explicit:

> The dagger has no obligation to take the parries in the same manner, as many practice, but … If the *tajo* or *reves* comes from high to low, the dagger is placed crossed, and if it comes from the side in the mode of a slap, the dagger is placed with the point straight to the sky in that place to where the blow goes. If it will be a *tajo*, one loads the dagger to his left side, and if it will be a *reves*, he loads it to the right side, which is not necessary when the *tajo* or *reves* comes from high to low. In the blows that will go to the legs, one will take the parries with the point toward the floor.
>
> (Godinho, 2016, p. 88)

Godinho here gives an easy to remember rule for the dagger: point to the side to intercept descending attacks (essentially, forming a roof), point up to intercept horizontal attacks (forming a wall), and point down to intercept low or ascending attacks (forming a floor or wall). The following play is a good example:

> Established nails-up with the daggers close to the swords, straight to the opponent, if one of them wants to lower the sword, he leaves his dagger as he had it. If the opponent travels to thrust nails-up to the chest, deflect it with the dagger and give a *reves* rounded over the head, into his head. The opponent can take this *reves* on the dagger with the point to the sky and give a nails-up thrust to the belly—

instead of deflecting and giving a *reves* as said above, the opponent can press his sword on it and descend with a nails-up thrust to the thigh, taking his dagger joined to the other's sword.

(Godinho, 2016, p. 84)

This play includes two dagger defenses, each displaying the pragmatism typical of escrima comun. In the beginning of the play, the fencer invites a thrust from the opponent, only to slap it away with the dagger and give a great backhanded cut to the head ("rounded" is one of Godinho's typical descriptions of a horizontal cut). Following his general rule, Godinho parries this cut with the dagger "with the point to the sky," crossing the dagger hand to the opponent's right, then thrusting low to the leg. As a result, the opponent's arms end up crossed. Bolognese is not entirely averse to crossing the arms in this manner, but tends to avoid it. A more typical Bolognese response to a backhanded cut like this would be to defend and counterattack with the sword, leaving the dagger out of the picture entirely.

Both schools of fencing were, in their own way, leery of using the dagger to offend the opponent when fencing with sword and dagger. The essential difficulty is that an opponent may also have a dagger, and medieval fencers had a healthy fear of knife fights. Godinho represents this attitude with an exception that proves the rule:

> He that plays against another dagger takes note that all his play is outside, because of the danger that exists with the daggers being close.
>
> When against a shield or buckler, one can choose that which suits him, choosing the inside or outside play, making use of both; one can take *tajos* or *reveses* on the sword, entering to wound with the dagger.

(Godinho, 2016, p. 83)

In other words, escrima comun is happy to close to knife range to strike with the dagger, but only if the opponent does not also have a dagger (it is a telling example of how much more historical fencers feared wounds with bladed implements than shield punches that Godinho recommends staying out of knife stabbing range but is happy to close to buckler punching range). Indeed, Godinho recommends that fencers without daggers entirely avoid cutting at an opponent armed with sword and dagger (i.e., that they stick to thrusts), lest their opponent close to dagger range under cover of the sword:

> [T]he shield, buckler, or cloak-wielder, having to battle a dagger wielder, in my advice, should avoid the blows of *tajo* and *reves*, because of the danger of the dagger wielder parrying them on the sword and entering to wound him with his dagger, without being able to be stopped.
>
> (Godinho, 2016, p. 92)

Perhaps because Bolognese texts tend to assume matched weapons for pedagogical purposes, Bolognese authors rarely recommend offending with the dagger at all. Those used to thinking of sword and dagger as an offensively-minded weapon combination (as opposed to the "defensive" choice of sword and shield or sword and cloak) may find this surprising. The wisdom of the advice to stay out of reach of an opponent's dagger is easily proven in sparring, however. The fact of the matter is that most of the time when one fencer is stabbed or cut by a dagger, so is the other one. Daggers—even very large ones with blades of 15" or more, such as seem to have been preferred by Bolognese fencers—are too small and too quick to easily control.

We have already seen this Bolognese principle in action against a forehand cut in the play from Manciolino. And, to be fair, it is fairly intuitive to defend with the dagger and attack with the sword when an attack comes from a fencer's dagger side. But what if the attack comes from a fencer's sword side? Godinho (assuming the opponent doesn't also have a dagger) would recommend that the fencer parry with the sword and hurry to close range to attack with the dagger. But the Bolognese advice is generally different. Dall'Agocchie considers this possibility in the following play. He begins with the sword held over the head with the point forward, on the fencer's right side, the dagger held low and on the right side, also with the point forward, with the right foot forward and the fencer's hips twisted to the right:

> But if he attacks you with a riverso to the head, draw your left foot near your right one and protect yourself in that tempo with your sword in the same guard, then immediately advance with

your right foot, thrusting your point into his chest and placing your dagger in defense of your head.

(dall'Agocchie, 2018, p. 51)

In this play, the fencer defends against a backhand cut to the head with the sword by advancing toward the cut with the left foot, so that the fencer's already upraised sword catches the incoming cut near the hilt and the fencer can immediately respond with a thrust, pushing the opponent's sword away with the hilt of the sword in the process. The dagger is a mere afterthought, raised above the head to protect against a second blow in case the opponent's first attack is a feint or the opponent is simply on the ball enough to throw a follow-up cut after being parried. The technique could theoretically be executed without the dagger at all—and, indeed, dall'Agocchie uses the daggerless version of this technique when discussing fencing with an unaccompanied sword.

Even in Bolognese, of course, the dagger may be used as a last-ditch defense against another dagger-armed opponent who unwisely closes to within dagger range, but deliberately closing to that range when both fencers have daggers is often a prescription for both fencers to end up stabbed. This is especially true given that each fencer must be aware not only of the opponent's dagger but also the opponent's sword. As a rule, wise fencers stay well away from an opponent's dagger regardless of system.

For this reason, sword and dagger fencing contains perhaps the least amount of grappling of any historical weapons combination. However, the dagger could still be used to trap the opponent's sword. Escrima comun contains a technique known as *encadenado* ("enchained") that is a prime example. The key to enchaining is using the dagger to rotate the opponent's sword around the defender's sword, controlling it with counterpressure exerted by both weapons. Here is Godinho's first play featuring this technique:

> Established nails-up with the dagger over the right arm ... if he comes with a nails-up thrust to the eyes, enchain him. How one enchains is that when the opponent comes with the thrust, you exchange the arms, passing the dagger arm to the right side, and the sword gives the nails-up thrust in the opponent's right shoulder,

and the moment said thrust goes, return the left arm to its side, pressing on the opponent's sword that is now locked between your dagger and sword.

(Godinho, 2016, p. 87)

In this play, the fencer begins with the palm facing up and the point of the sword forward, dagger held near the sword arm and pointing to the right so that the point projects past the sword arm. In response to an incoming thrust, the fencer thrusts back at the opponent's right shoulder, so that the opponent's sword comes in to the right of the defending fencer's sword. The dagger then sweeps left, out, and down, so that the defending fencer's sword acts as a fulcrum that the opponent's sword is forced to rotate around. This scissors or "enchains" the opponent's sword between sword and dagger, granting the enchaining fencer control of the opponent's weapon and the freedom to thrust home.

There is one last topic we should briefly mention before leaving sword and dagger. A common misconception about sword and dagger fencing is that the dagger can be used to push an opponent's sword out of the way before attacking. This may seem like an intuitive use for the dagger—after all, it keeps the sword, the longer weapon, free for offense—and it can work in a sparring environment, particularly against an inexperienced opponent. However, by the time a fencer is close enough to press on the opponent's sword with the dagger, the distance between the two is likely close enough that an experienced opponent can simply launch an attack. Perhaps for this reason, neither Bolognese nor escrima comun recommends this use of the dagger. This *is* a recommended use for a second sword, however, which brings us to the last topic for this chapter: the use of two swords.

Fencing with two swords at once ("dual wielding," as it is often called in pop culture) is a popular trope in sword-themed fantasy. It catches the eye, inspires the imagination … and is the subject of three misconceptions that we should address before we can discuss its historical methodology.

The first misconception is that European fencers did dual-wield, but the form was only for show or formalized duels. As we will discuss in this chapter, escrima comun contains extensive two sword material that can only apply to spontaneous street fighting.

The second misconception is that dual wielding was only performed with rapiers—that is, long, complex-hilted, thrust-oriented swords. It is true that some Renaissance authors discuss fencing with twin rapiers (a "case" of rapiers, as it was sometimes called). It is also true that both Bolognese and escrima comun discuss fencing with shorter, more cut-oriented weapons. The main hand sword in Marozzo's illustration for fencing with two swords, for instance, has only a single finger ring, making it hardly different from an arming sword. That said, it is easy to overstate how early dual wielding appears in the Middle Ages. As I argued briefly in Chapter 2, both Bolognese and escrima comun are medieval *systems*, but the only medieval *author* who discusses fencing with two swords is the 15th-century escrima comun master Pedro de la Torre. De la Torre wrote in 1474, and—according to Pacheco, at any rate—claims to have invented the Iberian two sword method. Whether or not that is true, 1474 is the earliest I can document dual wielding in any European fencing treatise. There is no evidence that fencers wielded two swords earlier in the Middle Ages, and no particular reason to believe that they did.

The third and final misconception is that dual wielding is exceptionally difficult. Historical authors do not speak of it as exceptionally difficult to learn, nor do they place it in their treatises in a position that might suggest they considered it exceptionally advanced, such as the final section of a book. Indeed, historical authors describe dual wielding as particularly *useful*, in direct contradiction to those who think of dual wielding as a mere show piece of skill. Manciolino calls it "quite advantageous and striking," while Marozzo calls it "a most excellent thing" (Manciolino, 2010, p. 126; Marozzo, 2018, p. 131).

With these misconceptions out of the way, we are ready to ask: what did Bolognese and escrima comun dual wielding look like?

Bolognese dual wielding, like Bolognese sword and dagger, is based on the "wheel-turning" principle of powering the movement of both weapons with a single turn of the hips. As with sword and dagger, one weapon can defend while the other attacks. With two swords, however, either weapon can be assigned either role. Consider the following play from Manciolino. The play begins with the fencer's right foot forward, with both swords toward the fencer's left side. The left sword is pointed high and to the left and the right sword is low and across the body with the point forward:

11. If the opponent ripostes with:
    a. A mandritto to your head delivered with his right sword, hit into his attack with your left sword, and deliver a thrust to his chest with your right sword.
    b. A riverso, parry it with your right sword, while striking his head with a mandritto to the face delivered with your left sword.

(Manciolino, 2010, p. 127)

In this play, the fencer defends with whichever sword is closest to the direction the attack is coming from. If the opponent throws a descending forehand cut (coming toward the fencer's left), the left sword parries while the right sword attacks. If the opponent throws a descending backhand cut (coming toward the fencer's right), the right sword parries and the left sword attacks. This level of interchangeability would not be possible with a sword and dagger, since the dagger would not have the reach to attack without closing quite precipitously. With two swords, it is easy.

Sidebar: Didn't swords get dull?

One of the questions I get asked most frequently is about edge retention and parries. Did historical fencers really parry with the edge of the sword (rather than, say, with the flat)? If they did, how did their swords stay sharp instead of getting saw-toothed and dull?

When I first discovered HEMA in the 1990s, there was a school of thought that held that all attacks could be parried with the flat of the blade. To be frank, this is no longer considered a valid theory by any reputable historical fencer.

The flat is used to parry occasionally, but flat parries with a cross-hilted sword have the distinct disadvantage that the crossguard is not oriented to protect the fingers or back of the hand. Thus, the flat can only reliably be used to parry when the blade is angled so that an incoming cut naturally slides toward the tip, away from the hands. When parrying with the edge, on the other hand, the crossguard is naturally oriented so that it intercepts the opponent's blade even if it slides toward the hand. This means that edge parries can be used to defend against any attack, regardless of angle, whereas flat parries cannot.

What is even more to the point for our purposes, however, is that as more treatises have been transcribed and translated since the 1990s, it has become absolutely clear that the majority of parries recommended by medieval systems are performed with the edge. This is not to say that swords were always slammed into each other at right angles, but it is to say that fencers parried more with their edges—at a variety of impact angles—than they did with their flats.

It is also to say that medieval fencers seem to have taken no particular care to "preserve" their edges in combat. Indeed, practical experimentation reveals that many historical techniques involving crossed swords are actually somewhat easier to perform with sharp blades and edge parries, as the sharp edges bite into each other a small amount and thus permit the point of contact to act more as a fulcrum than it does with blunt sparring weapons.

All of this raises the question of how blades did not become damaged and useless in combat. The answer to this question has several dimensions.

The first answer is that a sword can be made sharp enough to shave with without being razor-thin. There are, essentially, two ways to make a sword sharp: by polishing the cutting faces and by making the edge very thin. Judging by surviving antiques, the majority of medieval swords seem to have been sharpened primarily by polishing, leaving the edge supported by a healthy amount of metal. Thus, it simply is not the case that a shaving-sharp sword edge would crumple or chip at first contact with another edge, the rim of a shield, or armor.

The second answer, though, is that edges did indeed become damaged over time—and that this is not only acceptable, but inevitable. A fencer cannot always control what the sword's edge comes into contact with. The opponent will constantly be attempting to parry, after all, or at the very least to interpose body armor, and these efforts will inevitably result in a sword's edge coming into contact with things that will dull it over time. Even when this does not occur, the truth is that a sword's edge becomes less sharp to a small degree whenever it does absolutely anything. Even the act of cutting cloth, flesh, and bone—the very activity for which swords are sharpened in the first place—dulls an edge a small amount. Thus, the real answer to the question, "Didn't swords become dull and damaged over time?" is yes. Edge damage is inevitable and expected.

This is not necessarily catastrophic. Plenty of surviving antiques show evidence of such damage being ground out or otherwise repaired. How an edge is repaired is a question of how the damage occurred. There are, broadly speaking, three ways a sword edge can become dull. The most common is simply that the edge becomes less brightly polished over time. This can be fixed by re-polishing the sword, which is not an odious task so long as it is done on a regular basis. The next most common is for the very edge of the blade to roll over so that it is no longer in alignment with the rest of the sword. No steel has been lost, but it needs to be coaxed back into alignment. This can be done with a piece of steel or a strop, as is still done with straight razors today. Least commonly, a sword might chip, so that some of the steel is actually missing from the blade. This is the worst type of damage, and can only be repaired by re-grinding the edge to "blend" the hole back into the edge. Fortunately, it is also the least common.

Modern people sometimes have an obsession with edge preservation that medieval fencers do not seem to have shared. I suspect that this comes from the fact that we no longer sharpen blades on a regular basis in our daily lives. In fact, performing edge maintenance on a sword is something akin to cleaning a gun after it has been fired. It is normal maintenance, the expected result of operating the weapon, and trying to avoid it is simply silly.

The length of a second sword also permits it to harass the opponent, which is one of the signature advantages of a second sword over a companion dagger. Bolognese two sword fencing makes liberal use of this capability. For instance, the left sword can attack the opponent's sword hands, creating an opening that the right sword can exploit, as in this play from Marozzo:

> You remain with your left sword in cinghiara porta di ferro stretta, and your right sword in coda lunga e stretta, with your right leg forward. From here, hit your enemy's right hand with a rising falso with your left sword and step forward with your left foot, throwing a mandritto to his legs with your right sword.
>
> (Marozzo, 2018, p. 133)

Here, both swords begin on the fencer's right, with the right sword just outside the right knee and the left sword just inside it. The fencer steps forward, turning the hips to the left, and offers two attacks to the opponent: the left sword strikes with the back edge at the opponent's sword hand, and the right sword strikes at the opponent's leg with the front edge. Offering two attacks—one high, one low—makes it more likely that one of them will land. The attack to the hand serves almost as a distraction, allowing the leg cut to slip in while the opponent is busy defending the hand.

It is possible, of course, that the opponent will manage to defend both attacks (particularly if the opponent has two weapons as well). For this reason, another fundamental principle of Bolognese two sword fencing—and an important difference from Bolognese sword and dagger—is to use the length of one sword to control both of the opponent's, leaving the other sword a window of opportunity in which to strike unimpeded. Marozzo again, from the same starting position:

> [P]lace the true edge of your left sword on the outside of your enemy's left sword, that is, into its false edge.... When you put your true edge into his false edge, throw a falso impuntato to his left temple in the same tempo, stepping toward his left with your right foot.
>
> (Marozzo, 2018, p. 134)

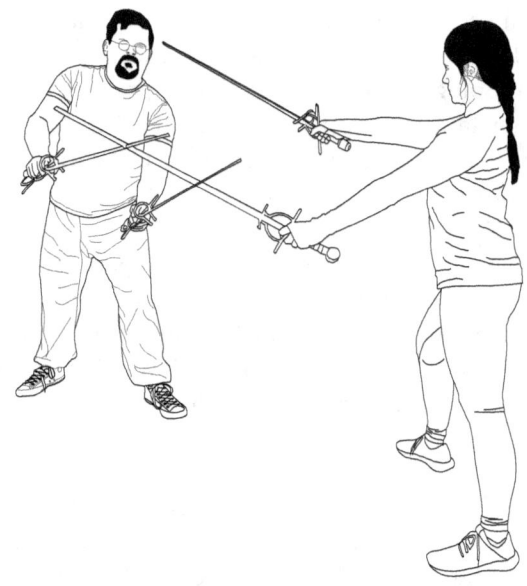

Again the fencer begins with both swords on the right. In this play, however, the left sword is placed over the opponent's left sword—and in practice, due to its length, over the opponent's right sword as well. With the left sword thus suppressing both of the opponent's weapons, the fencer steps around to the right and hooks a thrust into the left side of the opponent's head. As we have seen, the two weapons are interchangeable in practice; a similar maneuver could be performed starting from the fencer's left side, using the right sword to disable the opponent's weapons and the left sword to attack.

To sum up, Bolognese two sword material highlights three capabilities of two swords that no other weapon combination has: the ability to attack an opponent in two different places at once, the ability to use either hand interchangeably to attack or defend, and the ability to use the sword's length to suppress two weapons with one, freeing the second sword to attack unimpeded. Each of these capabilities is illustrated against a single opponent, which might lead one to conclude that dual wielding really was only useful for dueling or show. Escrima comun, however, has an entirely different focus.

From the beginning of his two sword material, Godinho assumes that the fencer armed with two swords is outnumbered and facing an unknown number of opponents in a variety of urban contexts, which

THE FORGOTTEN 121

Godinho delineates as streets of various sizes. He begins his two swords section with a complicated cutting pattern that it is worth looking at in some detail, suitable for "a street that is not very narrow":

> Drawing the two swords, the one in the left hand is withdrawn below the right arm, and the one in the right hand will be placed nails-down above the head. Putting in the left foot, throw a *reves* with the left arm and a *tajo* with the right arm at the same time—these two blows end with the left sword above the head nails-down, and the sword in the right hand below the one in the left. Then, put in the right foot with a *reves* of the right hand and a *tajo* with the left arm, the swords ending in the first posture, which is the left arm below the right, and the right nails-down above the head. These two steps done, return to do others like them, continuing the two blows of *tajo* and *reves* in said manner, always responding to each step of the foot with a *tajo* and *reves*, the swords always ending as it is said in the two steps, and during the battle, continue the step and blows without change.
>
> (Godinho, 2016, p. 73)

In this pattern (or "rule," as Godinho calls it), the fencer begins with the right hand sword pointing forward, extended at roughly eye height with the palm facing down, while the left sword is tucked under the left armpit facing behind, also with the palm facing down. The fencer makes either an ascending or horizontal backhanded cut with the left sword, following with a diagonal descending forehand cut with the right sword as soon as the left arm's cut has carried it out of the path of the right sword. As the right hand sword settles under the left armpit, the left hand sword circles around the head to make another horizontal backhanded cut, ending facing forward at eye height. The pattern then repeats in mirror image from the left side.

The result of this rule is that the fencer is surrounded by cuts that travel around the body one after the other in very rapid succession. This obviates the greatest difficulty of facing multiple opponents with a sword, which is the risk that opponents will rush the outnumbered fencer as soon as their victim's sword has passed by them. Escrima comun's two sword method is founded on using the two swords to throw cuts so quickly that opponents never have such an opening. Notice in the rule above, for instance, that the basic three cut pattern includes two cuts from right to left and then a third that circles around the head again, filling the space on the fencer's right side just as an opponent would think it was clear.

The basic two sword rule is, as its chapter heading indicates, best suited to situations where the fencer's attackers are all to the front. The fifth rule provides Godinho's rule for a dual wielder who is actually surrounded:

> The two swords being surrounded, throw a *tajo* in a wheel, arming the wheel on the left foot, and finish it with the same foot ending directly to the other side, in a manner that if your face was to one side when you committed the *tajo*, it will be to rear [sic] when finishing the *tajo*. Then throw another *tajo* in a wheel, arming on the right foot, with the left foot ending in front in a manner that the face ends up toward the other side, as said above. In this manner, you will go with this step, playing two, three, or four *tajos*, however many will be necessary for the space that the opponents will enclose. The *tajos* finished, the left arm leaves with a *reves*, and then turn with the same *tajos* in the manner above to that place where you had left, keeping to the step in all.
>
> (Godinho, 2016, p. 76)

This rule has some similarities to the first one we looked at, with its wide cuts that encircle the body. The footwork is noticeably different, however. In this rule the dual wielder turns 180 degrees with every step, essentially pivoting across the available space while throwing wide circular cuts to keep the opponents at bay. To modern eyes, the pivoting footwork may seem needlessly cinematic. Godinho, of course, was not imitating Hollywood. Rather, the pivots in this rule help extend the reach of each cut and sweep the fencer's eyes across the entire area in which the fight is taking place. Both of these are crucial considerations when a fencer is not merely outnumbered but actually surrounded.

This rule neatly illustrates the importance of keeping the objective in mind when fencing outnumbered. As Godinho indicates, a surrounded fencer should throw "two, three, or four" pivoting forehand cuts, progressing in a more or less linear direction the whole time. The fencer's objective is not to cut down each opponent, but to keep them at bay beyond the circle of the two swords until a break in the attackers' cordon appears. Once it does, the fencer will escape through it and withdraw.

Precisely because cutting one's way out of a street fight may not leave behind a bloody pile of bodies (or, indeed, any bodies at all), Godinho includes a special rule for withdrawing:

> [M]ake a step from one side to the other, namely to those of the left side, and to those of the right side, never turning the back to one side or the other, but the body always straight, [and] play on one side taking the feet to one side, one behind another, throwing a *reves* with the left arm and a *tajo* with the right arm to those on the right side. The *reves* ends with a nails-down punching thrust to those that remain on the left side, and the sword of the *tajo* goes below the left arm.
>
> (Godinho, 2016, p. 80–81)

Godinho then mirrors this rule against those opponents on the left side, "keeping to the rule of the step and blows to one side and to another, as long as the fight and posture will last" (Godinho, 2016, p. 80).

The withdrawing rule is very similar to the first rule for a somewhat narrow street, which is only fitting since in both situations the attackers are on one side of the fencer but may be seeking to spread out again. The main differences are that the fencer is stepping backward and the addition of the punching thrusts to further deter the opponents from pursuing too closely (punching thrusts, which Godinho notes elsewhere are especially powerful, begin with the hand withdrawn before "punching" forward).

You may be wondering how a fight like this actually ends, if the dual wielder is not necessarily expected to kill all (or any) of the attackers. Killing or wounding all of the attackers is a possible outcome, of course, but Godinho's real endgame is indicated by his last comment: "as long as the fight and posture will last." His meaning seems to be that, if the defending fencer can hold the opponents off long enough while retreating, at some point the fight will simply peter out, or be broken up by bystanders. The presence of bystanders who could call for help or come to the outnumbered fencer's aid is implied by the very titles of the rules: if one is fighting in a "street," then other people must be nearby. This is another important context clue for determining that Godinho intends his two sword material not for the dueling salon but for spontaneous street defense. Indeed, his tenth rule for two swords is entitled "separating a fight," which is worth quoting at length:

> When it will be necessary to come to break up a fight, take note that the blows that one plays are all armed and played above the head, and on no occasion is there a thrust, so that in the brawl, it doesn't run into anyone, nor are the *tajos*, and *reveses* so low that they hit anyone, only flown above the head, intending to place yourself in the middle. Being armed on one of the feet, go to one side in a continuous motion, giving leaps with the feet while playing *reveses*. Turning in the same manner with the same *reveses* with a single sword, but when giving the turn, at the time that he turns, he plays a *tajo* and *reves* with both the swords until he turns to part. Not wanting to start off with *reveses*, start off with *tajos*, but in said manner. The sword that isn't cutting goes withdrawn below the arm of the other the moment it's not cutting. Take note that more resistance is given to the side that is stronger.
>
> (Godinho, 2016, p. 80)

Here, the fencer uses the two swords' ability to encircle the body with rapid cuts to interpose within a "brawl," moving back and forth to separate the two sides. The basic motion is similar to fighting multiple opponents, with the important note that the fencer is deliberately throwing cuts high and avoiding thrusts so as not to injure any of the combatants (injuring the combatants you are trying to separate, you can probably imagine, might well make the would-be peacemaker a target). Godinho's concern about injuring opponents in this rule is also indirect confirmation that the cuts and thrusts in his other rules are intended to wound if an attacker is foolish enough to get that close.

It is doubtful that most Iberian fencers who broke up brawls happened to be wearing two swords on them when they did so. However, this rule is a good example of the obligation of bystanders to break up fights when possible. It is presumably this intervention by bystanders—whether with swords or simply the social pressure of crowds—that Godinho is counting on to end most fights against multiple opponents.

In all, Godinho includes 11 different rules for fencing with two swords in different street fighting contexts, including when a fencer's opponents are all on one side, before and behind, or all around; in streets ranging from "very narrow" to "wide"; with objectives including cutting free of the attackers, guarding a lady or goods from ruffians, and breaking up a street fight. Each rule has a slightly different pattern of footwork and cuts and thrusts, but all share the basic idea of the first rule of surrounding the fencer with a defensive shield of cuts and thrusts that come too fast for opponents to find an opening in the pattern that they can exploit.

It is worth noting that these rules are not intended to be adhered to religiously. Rather, the fencer is expected to flow freely from one rule to another as the situation demands. Godinho refers to this idea late in his third rule, which discusses the scenario of a fencer attacked by opponents who have spread out but are still all on one side. "If surrounded," he says, "leave it and do the rule of being surrounded, and if retreating, leave the one and take the other" (Godinho, 2016, p. 74). The rules are tools to help the fencer understand how to counter the various threat axes of different scenarios, not talismans to execute without any critical thought.

We shall end our discussion of two swords by looking at Godinho's ninth rule, entitled "four streets," which exemplifies many of Godinho's principles. In this scenario, a dual wielding fencer is caught at a crossroads or plaza, with opponents on all sides (from the number of scenarios Godinho discusses in which a fencer faces an apparently premeditated attack by multiple opponents, it is difficult to avoid the conclusion that Godinho traveled in dangerous circles). He writes:

> [S]tanding firm and bending the knees, play a *tajo* and *reves* in a manner that if possible, these two blows surround the whole body, and then like this, from standing firm, another *tajo* and *reves* in the same manner as those from above. Turning in a continuous motion on the left foot, leave with a *reves*, giving a jump to one of the corners. Stopping, arm another two steps like those above.... Each step is made like I say: with two *reveses* and two *tajos* standing firm, leave with the *reveses* turned above the head, and while falling with the jump, quickly arm the step of the two *reveses* and two *tajos*, taking one of the corners at each jump. If it will be necessary to give a jump in the middle of the street, do it; it is not bad, because the opponents do not press themselves together in a short and tight ring, but it is best of all to do one of the jumps of the step to gain one of the streets, and then do the rule that suits.
>
> (Godinho, 2016, pp. 79–80)

When Godinho says "jump," he means it. In a situation this dire, the fencer must be exceptionally active to survive. Above all, the fencer must prevent the opponents from forming a "short and tight ring," at which point no amount of fancy sword work will save their victim from being cut to pieces. To avoid this, the fencer literally leaps from one threat source to another, whirling both swords around the whole body at each stop to keep the opponents at bay. The fencer maintains this movement as long as necessary, keeping the opponents' cordon open and loose. As the opponents jockey for position, the fencer keeps one eye on the surrounding streets, leaping toward one as soon as possible and then withdrawing from the attack. This rule combines Godinho's shielding patterns of cuts with unpredictable movement, and also demonstrates his focus on escape rather than killing the attackers.

Modern observers of fencing like this often wonder when the multiple attackers are actually struck. The answer is that they are only struck incidentally. An unwary attacker can certainly catch a sword to the face thanks to the difficulty of predicting when a skilled escrima comun fencer has his attention somewhere else, but from the perspective of the dual wielder, it is perfectly acceptable for none of the attackers to die. The goal of this sort of street fighting is not to cut down every single attacker, leaving the outnumbered fencer victorious on the blood-soaked streets. The goal is escape: as Godinho says, it is "best of all" for the fencer's seemingly random jumps within the ring of attackers to subtly press on one avenue so that the fencer can narrow the number of threats that must be dealt with and, ultimately, flee.

> **Sidebar: Which would win: sword and buckler or two swords?**
>
> In this chapter we looked at three companion weapons that eventually replaced the buckler as the standard companion weapon to the one-handed sword: the cloak, the dagger, and the second sword. Both cloak and dagger have some unique capabilities compared to a buckler, but as a general rule, they are inferior companion weapons. Two swords, however, present an interesting comparison to sword and buckler. Was sword and buckler superior to dual wielding? And if it was not, why wasn't two swords the default medieval civilian weapon combination, instead of sword and buckler?

I'm aware of no historical sources that discuss this matchup directly (sword and buckler is on its last legs by the time two swords appears on the scene; as we discussed, it's something of a stretch to call two swords a "medieval" weapon combination at all). However, we can extrapolate based on what treatises say about each weapon form and the modern experience of pitting them against each other.

So far as modern experience goes, two swords generally seems to have the advantage over sword and buckler. For a fencer armed with sword and buckler attacking a fencer armed with two swords, the two swords present largely the same problems as would another sword and buckler: the opponent still has two weapons, and the sword and buckler fencer must overcome both before it is safe to land a blow. However, when the sword and buckler fencer is defending, two swords are not at all equivalent to a sword and buckler, for the second sword has a longer reach and greater wounding potential than does a buckler. Essentially, a sword and buckler fencer is inherently asymmetrical, while a fencer armed with two swords is a symmetrical opponent. This makes it much easier to overwhelm the defenses of a sword and buckler: by throwing multiple attacks against the sword and buckler fencer's left or right side, for instance, or by repeatedly striking low, where the buckler cannot easily reach.

However, we should consider what our treatises have to say, as well. Bucklers are, after all, a very powerful defensive implement. Were they considered a more powerful defensive implement than a second full-sized sword?

As we saw in Chapter 4, bucklers can really only do four things. They can intercept weapon strikes from roughly head to waist height, including strikes against the sword hand. They can passively close off lines of attack, primarily (though not, as Godinho would remind us, exclusively) on the shield side. They can disable an opponent's weapon by pushing on it (for instance, into the opponent's chest). And they can injure the opponent by punching, mostly in the face.

Looked at this way, a second sword does everything a buckler can do equally or better: it can intercept strikes from head to toe, passively close off lines of attack (and a fencer with two swords can close off lines on either side of the body equally well), disable the opponent's weapon by trapping or manipulating it, and injure the opponent in a variety of ways in a variety of places.

In addition to all that, two swords can do things that a sword and buckler cannot. They can threaten two targets at once, including in multiple directions (as we have seen, holding off multiple opponents was a particular specialty of two swords). A second sword is also more versatile when it comes to overcoming an opponent's defenses. When a buckler meets another weapon, all it can really do is drive hard (pushing the other weapon around). A sword can do that too, but it can also yield around to offend the opponent, such as with one of the wrist cuts we have seen one-handed swords make such use of, or by angling a thrust to slide around the defending weapon.

So, do bucklers have a significant defensive edge (pun intended) over a second sword? Not really.

Why then was the sword and buckler so popular in the Middle Ages (and, to a decreasing degree, in the Renaissance)? Why was two swords not more popular as a civilian weapon combination?

The answer to this question is a mystery. No fencing master discusses it to my knowledge, not even during the Renaissance, when dual wielding was at its most popular.

One potential answer that I hear a lot is that two swords are just harder to learn than sword and buckler. This is an explanation with which I can only partly agree. Using two swords is certainly more *tiring*, at least in my experience. However, I am not convinced by arguments that two swords are inherently harder to learn or use than a sword and buckler. For one thing, the difficulty of using a sword and buckler depends to a significant degree on the system and tactical paradigm being used. For instance, it is easier for most people to wrap your head around using a buckler as a detachable hand guard than it is to master using the buckler in a more independent manner; in the same way, you can treat your second sword as a superior parrying dagger, which is a lot easier than treating the two swords as actually interchangeable. It's also worth noting that treatises that discuss dual wielding do not warn the reader about how hard it is to learn, or otherwise give any indication that it was thought of in period as an especially advanced skill. While it is *possible* to learn to use sword and buckler in a way that is easier than *some* ways to use two swords, this seems like an incomplete explanation at best.

A second piece of the puzzle, and perhaps the largest piece, is found in the socio-legal implications of civilian-carried weapons. The Middle

Ages was not a computer roleplaying game, in which characters can walk about town armed to the teeth and provoke no reaction from those around them. Two swords may simply have provoked too negative a reaction for it to be very popular as a civilian weapon combination—particularly if we recall two swords' noted facility for dealing with multiple opponents. There is a fine line between appearing respectably armed for self-defense and appearing eager to take on a mob of marketplace toughs at a moment's notice.

We should pause here to deal with a potential objection to this line of speculation. The reader may object that if two swords are really more capable than sword and buckler, it makes little sense for two swords to *replace* sword and buckler as an "acceptable" weapon combinations. There is some justice to this complaint. At the same time, it is important to remember that social attitudes toward acceptable civilian armament do not progress linearly, or in ways that always seem rational from the remove of many centuries. To take an example from the world of firearms, we might ask whether it is reasonable for a law-abiding, peaceably-minded man to carry a pistol for self-defense. Depending on your time and place, the answer may be no (in many places, wheellock pistols were banned when first introduced) or yes (many American jurisdictions permit people to carry self-loading pistols today, which would seem like a ludicrous amount of firepower to most 16th century people) or no (many European jurisdictions do not permit people to carry any kind of pistol today, which seems like a ludicrously small amount of firepower to many 21st century Americans). Sociolegal rules about weapons fluctuate. It may have been that two swords benefited from just such a fluctuation, which barely (and briefly) brought them into the realm of respectability.

It must also be said that, while dual wielding was never *restricted* to use in duels, it was *used* in duels (as we have seen, this seems to be the primary use case envisioned by the Bolognese). The changing face of dueling may thus have also had an influence on the popularity of the practice. For instance, two swords are of little use against armor, and in an era when the most prestigious duels were fought in full knightly kit, dual wielding may have lacked a certain cachet. Later, when authors like dall'Agocchie could write admiringly of the duelist who dared to fight without armor, this may have changed.

The last thing to remember about a sword and buckler is the ease of carrying them. A buckler can be hung on the scabbard, which puts the entire weapon combination on one side of the body. This is reasonably convenient: the fencer can angle to one side when navigating crowded spaces, and it only requires one hand to "steer" the weapons or stop them from banging uncomfortably against the leg. Two swords must be worn on both sides of the body, which makes it harder to navigate crowded spaces while wearing them and may require a fencer to use both hands to steer through a crowd. Thus, while two swords may be a superior weapon combination for offense and defense, a sword and buckler is significantly more accommodating to everyday life. Perhaps we should not be surprised, then, that it is the sword and buckler that Marozzo praises as "excellent and useful" (Marozzo, 2018, p. 136).

CHAPTER SEVEN

Malice and art: knives and daggers

> *I'm the noble weapon named the dagger and I play at very close range. He who understands my malice and my art will also gain a good understanding of many other weapons. I finish my fight so fiercely and quickly that there's no man who can stand against my method. If you watch my deeds of arms you'll see me make covers and thrusts as I move to grapple, and take away the dagger by dislocating and bending arms. Neither weapons nor armor will be of any use against me.*
> —Fiore dei Liberi (Fiore, 2017, p. 9v)

Daggers are fun. Dagger *fencing* is fun. It is possibly my favorite weapon to fight with in terms of sheer enjoyment. A dagger duel with sparring daggers is fast, highly technical, and athletic. It is also among the more artificial activities in HEMA.

One of the criticisms leveled against historical fencing is that the arts found in surviving fencing manuals are too focused on individual combat, the dueling context of matched weapons, or both to have been of use in the "real world" of a battlefield or spontaneous self-defense scenario. We'll discuss the battlefield implications of historical fencing treatises in Chapter 13, and we have already seen numerous cases that put the lie to the idea that those treatises or their authors knew nothing of "real" combat. This is not to say, though, that historical fencing masters were *unaware* of the difference between fencing in earnest and fencing for practice, fun, or exhibition. Few surviving bodies of work highlight this difference more starkly than surviving advice about dagger fencing.

Medieval dagger work can be divided into three categories. The largest portion of surviving material focuses on unarmored (and often unarmed) defense against a dagger attack—what I will call "dagger defense" in this chapter. The next largest portion concerns the use of a dagger against an opponent in armor. Lastly, Marozzo contains a small amount of material on the use of a dagger in a formal duel in which both combatants are armed with nothing but daggers, or daggers and cloaks. We will discuss armored dagger fencing—such as it is—in Chapter 12. In this chapter, we will look at unarmored medieval dagger fencing.

With the exception of Marozzo's dagger duel, the type of dagger used in this chapter is not important. HEMA fencers doing dagger work tend to default to rondel daggers, which sandwich the dagger hand between large circular hand guards that allow a fencer to concentrate the maximum amount of force into a dagger stab (rondel daggers are in origin a military weapon, designed for the difficult task of stabbing through mail armor, but that did not stop many "civilian" weapons from sporting the hilt type). However, many dagger treatises do not specify the type of dagger they have in mind, and Marozzo illustrates a wide variety in his book. When daggers are illustrated, they tend to have very long blades compared to those used in most modern knife attacks. Blades as long as a forearm are common. It is worth remembering that medieval people were much more likely than modern people to be able to wear such large blades without arousing the suspicion of those around them.

Dagger defense is treated by KDF, armizare, Bolognese, and escrima comun. Despite this, there is broad agreement across all four systems as to the principles of how a fencer should respond to a spontaneous dagger attack. Fiore sums up the agreed-upon advice:

> Everyone should be careful when fencing the perilous dagger, and you should move quickly against it with your arms, hands and elbows, to do these five things, namely: take away the dagger;

strike; dislocate the arms; bind the arms; and force your opponent to the ground. You should never fail to do one or the other of these five things.

(Fiore, 2017, p. 9v)

Notably missing from Fiore's list is the sort of thrust and parry that the word "fencing" typically calls to mind. Indeed, medieval dagger defense is fairly characterized as wrestling with knives. The fact that one is armed with a dagger is almost an afterthought.

It is easy, especially in a modern sparring context where the blades are not real, for fencers to become overly focused on using their dagger. The usual result of this is an orgy of mutual stabbing. By contrast, the focus of our surviving historical material is always on the *opponent's* dagger. Lignitzer shows this focus succinctly in his first dagger technique:

> If he thrusts down from above to your face or breast, then go up from below with your left arm and catch the thrust on your arm. And grasp with your left hand from inside out over his right arm and press it fast to your left side. And then thrust with your dagger to his face.

(Von Danzig, 2010, p. 171)

Lignitzer has the fencer first catch a downward stab (the attacker's dagger is held in an "icepick" grip, with the blade extending below the attacker's pinkie finger) with the left arm. In practice, the inside of the defending fencer's wrist should contact the inside of the attacker's wrist. This stops the incoming thrust and permits the defender to wrap the left hand over the attacker's forearm and crank it down, turning the attacker's hand over in the process so that the dagger points to the sky. This action presses against the dagger blade with the defender's forearm, keeping it from moving, as the defender pins it against the body (or, if the motion is sufficiently forceful, popping it out of the attacker's hand entirely). Only *after* the attacker's dagger is thus trapped does Lignitzer advise the defender to strike.

In practice, this all happens in the blink of an eye, but the sequence of events is nevertheless important. Dagger fighting is notoriously chaotic, and all of the defender's focus must be on securing or otherwise neutralizing the opponent's dagger—not striking. Lignitzer's defense is a common one (it is also Fiore's first dagger defense). If performed well, it traps the attacker's blade against the forearm tightly enough that the attacker loses effective control of the weapon. At the blink-and-you-missed-it speed of a dagger fight, however, defenses are often not performed perfectly. Martin Huntzfelt, another KDF master and possible contemporary of Lignitzer, supplies the obvious counter to Lignitzer's first technique:

> *Note:* when you thrust to an opponent down from above and hold the dagger with the hand guard behind your hand, and he parries with his hand reversed—it can be either hand—and means to twist your right arm around or force out your dagger, then wind your dagger over his hand, and slice off against his hand.
> (Von Danzig, 2010, p. 180)

If the attacker's wrist is free to manipulate the dagger, it can be sliced across the defender's wrist, which is precisely what Huntzfelt recommends here. The dagger can also be driven into the defender's chest instead. Fiore includes the latter counter, writing:

> I'll turn my dagger around your arm. And because of this counter you won't be able to take the dagger from me. Also with this turn I'll drive my dagger into your chest for certain.
> (Fiore, 2017, p. 10v)

A good sense of the high-stakes wrestling game of medieval dagger defense can be had from following Fiore's counters and counter-counters to this first, basic defense. As already noted, it is common for any dagger defense to be executed imperfectly in the heat of the moment. One common error is for the defender to miss the wrist-to-wrist contact that makes Lignitzer's and Fiore's first defense work, striking the attacker's forearm instead of the wrist. In this scenario (assuming the failed defense was still adequate to stop the incoming thrust), Fiore recommends diving the left arm over and under the attacker's dagger arm and wrenching upwards on the elbow while turning to the right. This turns the attacker's arm like a key in a lock, ending with the arm trapped behind the defender's back. From this position, the defender can drop to the ground, dragging the attacker down as well. Fiore writes:

> I'll lock your arm in the middle bind [la mezana ligadura], and I'll do it in such a way that you won't be able to give me any trouble.
>
> (Fiore, 2017, p. 10v)

However, the attacker can grab the dagger and use it to crank the attacker forward by putting pressure on the shoulder joint:

> I make the counter to the play that came before me. You can see the kind of position that I've put him in. From here I'll break his arm or quickly throw him to the ground.
>
> (Fiore, 2017, p. 10v)

From start to finish, the sequence involves a stab from the attacker, which the defender counters by trapping the attacker's dagger arm with the defender's left, only to be bent over face first and forced to the ground when the attacker's left arm comes into play to use the arm trap as offensive leverage. It is, in other words, a mad scramble of desperate wrestling for control of the dagger, which could well end in a broken arm for one or the other combatant.

Even when both fencers have drawn daggers, the standard medieval method of dagger defense was grappling-centric. Even though medieval dagger blades tended to be quite long (again, blades of forearm length or longer do not seem to have been uncommon, judging by illustrations), parrying daggers blade-to-blade is exceptionally difficult. This is a

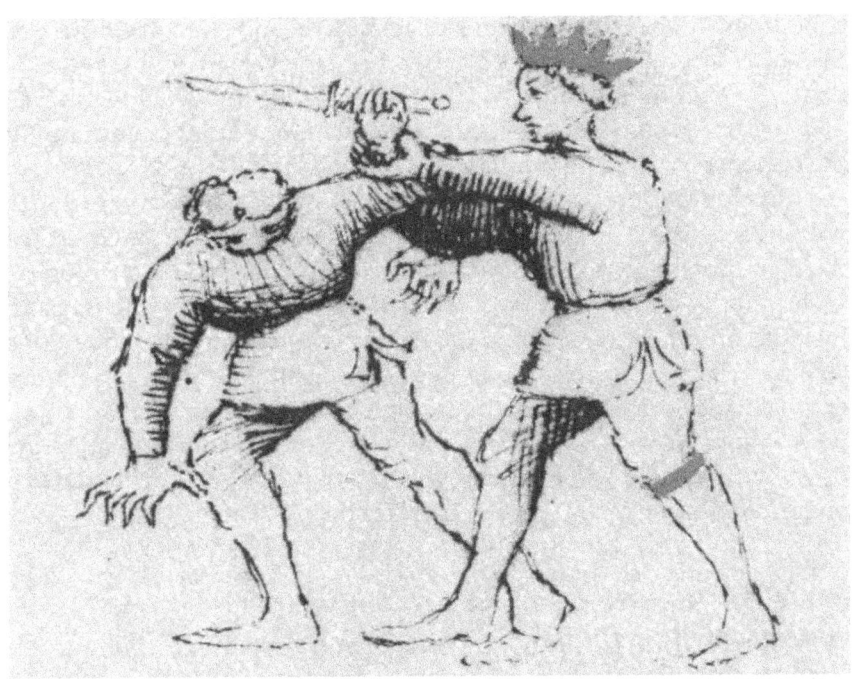

significant difference between historical techniques for sword and dagger. Monte concurs, writing of dagger fighting that "when the fighters remain close and have short weapons, it is not so easy to parry" (Monte, 2018).

Thus, rather than a parry-riposte model, dagger vs. dagger techniques tend to use the dagger merely as an aid to stopping the opponent's attack before initiating an otherwise standard grappling or disarming technique. Huntzfelt gives a good example of this, which he calls "wrenching" the dagger:

> [W]hen he holds his dagger with the hand guard in front on his hand, then hold yours the same way. If he thrusts at you below, then thrust at the same time and bind his dagger with yours. And grasp his dagger quickly down from above with your left hand so that your thumb stands toward you. And turn your dagger to his right hand, and with your left hand break his dagger upwards over your dagger. Thus you take his dagger.
> (Von Danzig, 2010, p. 180)

Here, the defending fencer stops an upward thrust with a dagger parry at an angle so that it forms a fulcrum over the back of the opponent's hand, then seizes the opponent's dagger blade and cranks it back over the fulcrum of the parrying dagger, ripping it from the opponent's hand. Despite the fact that the play begins with the defender having already drawn a dagger, the focus of the technique is still on disarming the opponent through grappling.

Huntzfelt's approach is by no means anomalous. Marozzo includes only two defenses in which both attacker and defender have drawn daggers, but both of them advise disabling the opponent's dagger with grappling before the defender's dagger strikes any blows. Fiore includes three entire sections on dagger vs. dagger techniques (his sixth, seventh, and eighth dagger masters), all but one of the techniques of which are dedicated to methods of grappling a dagger-armed opponent with one's own dagger. The one exception is a counter-stab to the opponent's dagger hand, a focus which is similar to Marozzo's advice on formal dagger vs. dagger dueling (see below).

> **Sidebar: Just how good at wrestling were historical dagger fighters?**
>
> In a modern competitive context, historical fencing matches with daggers are often either tentative affairs in which fencers snipe at each other without wrestling at all or brutish affairs in which fencers simply charge each other in a frenzy of stabbing, also without any wrestling. While both of these scenarios have a certain realism to them, it can leave the modern observer wondering how realistic historical dagger defense techniques are, with their emphasis on grappling and disarms. You may be wondering the same thing, just from reading this chapter so far.
>
> It is always a mistake to expect actual fencing to "look like the books" in a strict sense. The chaos of unscripted fencing lends itself to the application of sound principles from historical techniques, but it is rare to encounter the exact circumstances required to apply the precise "textbook" version of a play. Nonetheless, one of the great limitations of modern historical dagger fencing is that most modern fencers have substantially less wrestling experience than our historical counterparts.
>
> The level of grappling capability historical authors assumed can be inferred from some of their plays. Marozzo's 19th *presa* (grappling technique) against a dagger is a sacrifice throw, in which the defending fencer falls backward to the ground and pitches the attacker over the defender, onto the attacker's head:

[W]hen he approaches you with his weapon underhand to kill you, or inflict some wounds ... save yourself by grabbing his right arm with your left hand, taking it upward. Grab his left arm with your right hand, holding it strongly and tightly, and immediately upon grasping him, let yourself fall backwards to the ground, putting both your feet in his belly or chest as you fall, drawing in your arms, and throw him backwards over your head using your feet.

(Marozzo, 2018, p. 331)

Lignitzer's seventh technique is a hip throw:

If he draws his dagger before you draw yours and thrusts at you above, then catch the thrust on your left arm and go from inside out over his right arm. And press to your left side and swing away from him to your right side ... spring with your right leg to your left side, and grasp with your right hand under his right arm; thus you throw him over your hip.

(Von Danzig, 2010, p. 172)

The number of throws involved in medieval dagger defense is a good indication that these techniques were intended for accomplished grapplers. This is not a surprise when one considers that a goodly portion of armored fencing involves grappling at one point or another (for more on which, see Chapter 12). For a medieval fighting man, grappling was serious business, and it would only make sense to apply those skills as broadly as possible. Fiore illustrates the broad applicability of grappling skills in the third remedy to his third dagger master. In this scenario, the defender has secured a grip around the opponent's neck and stepped behind the opponent so their right legs are touching. By tilting the opponent's head, the defender throws the attacker backward over the right leg. Fiore writes:

> With this play I'll drive you into the ground, and in armor I couldn't make a safer throw. But even if I'm not in armor, there's still nothing you can do, and even if you were very strong and powerful, I'd still be able to do this to you.
>
> (Fiore, 2017, p. 13v)

The existence of such complicated wrestling plays should not be taken as a claim that all knife attacks occurred between expert grapplers, or even that expert grapplers could necessarily bring their skills into play during a dagger attack. On the contrary, medieval knife attacks (as modern knife attacks) could come with so little warning that defense was essentially impossible. Godinho is unusually frank about this among fencing masters by giving advice about how best to start a fight. For instance, he recommends the following method of initiating a dagger attack:

> If you want, being close at the end of the conversation, before you draw, you can give him a slap or throw a hand to the quillons of his sword, and then draw the dagger ... Also, you can throw a hand to the brim of his hat and press on it to put it over his eyes. In all this, be wary that he doesn't carry a dagger with which he offends you at the same time, or when carrying it, do it cautiously as you must.
> (Godinho, 2016, p. 117–118)

Note that the attack is initiated "at the end of the conversation." Essentially, Godinho's advice is to ambush the victim when he least expects it, at point-blank range, in such a way as to disorient him or prevent him from withdrawing to make space or drawing his sword—and, preferably, to do all of this to a victim who isn't wearing a dagger of his own. In such circumstances, the chances of even a highly skilled fencer surviving the dagger attack are remote. Godinho includes an entire section of his book of such "dirty tricks" for initiating a fight, including such tricks as swearing to an opponent on the cross that you have no further quarrel with him (only to poke him in the eyes and draw your weapon), kicking out his knee, and hurling pocket sand (which, he notes, you should carry in *both* pockets). Tellingly, all of these suggestions are listed under the heading, "Against Treachery," which suggests that Godinho's real advice is to expect an opponent to use such dirty tricks on you if you don't do it first.

It would, however, be a mistake to characterize historical dagger work as consisting solely of treacherous turbo-murder against which there is no defense. As we saw as far back as Chapter 3, armed medieval violence existed on a spectrum of lethal intent that modern people often have a hard time imagining. In the 21st century, an attacker drawing a knife (let alone a sword!) is likely to be taken as definitive proof of murderous intent, which in turn is likely to justify a lethal response. In the 15th and 16th centuries, the picture was not always so clear-cut, and historical fencers needed a proportionate range of lethal and non-lethal responses.

Marozzo's dagger defense section includes three scenarios in which a dagger-armed opponent attacks with raised dagger after first grabbing his victim by the shirt with his other hand. The hand on the chest suggests that the attack comes not as the result of a premeditated ambush killing but an escalating argument (or, perhaps, a robbery). Marozzo's 16th presa against a dagger is one such:

> Now take note that if your enemy were to grab you by the chest with his left hand to wound you with his dagger, overhand, you would extricate yourself by driving both of your arms together down onto his left arm. By doing so you will emerge safe from harm.
>
> (Marozzo, 2018, p. 325)

It is especially interesting to note that this defense simply results in the defender breaking away (contrast with Godinho's advice, which boils down to "stab him to death before he can even lay a hand on you"). While the defender could certainly draw a weapon at this point to continue the fight, it is also not hard to imagine him attempting to de-escalate the situation or simply running away. Certainly, if the defender does manage to draw his own dagger after breaking away, the attacker would be well advised to halt his attack. As Godinho's advice suggests, the best victim of a dagger attack is one without a dagger of his own. Manciolino concurs about the chaotic mess of a fight with daggers:

> The shorter the weapons, the more dangerous, since those which can strike from close-by are of greater risk, as their blows cannot be easily parried due to the speed at which they arrive. In consequence, the partisan carries more danger than the spear, and the dagger more than the sword.
>
> (Manciolino, 2010, p. 77)

Paradoxically, it is probably this very danger that led Marozzo to include his short sections on dagger dueling. A dagger duel, he says, is a "very useful thing for the one who has to choose the weapons"; i.e., a challenged party who demands a duel with daggers may cause his challenger to reconsider (Marozzo, 2018, p. 120).

Marozzo's advice for dagger dueling is founded on the principle of attacking the opponent's dagger arm, making attacks to the head or body only when the fencer has an absolutely clear window in which to do so. In this, he is in broad agreement with the other systems we have looked at, which focus on controlling or disabling the opponent's dagger or dagger hand. The first section of his dagger duel shows a number of familiar principles:

> Don't ever take your eye off your enemy's dagger hand, so that you can begin by throwing a thrust to the inside of his right hand, to the part that his glove doesn't cover; and if he doesn't have a glove, so much the worse for him. Along with that thrust, turn a mezzo mandritto the same way, so that your dagger will go into porta di ferro alta. If your enemy then throws a blow to your head, step a bit to his right with your right foot, taking his blow on the true edge of your dagger in guardia di intrare as you step. In that same parry, step to his right side with your left foot and grab his right arm with your left hand, moving it outwards so that you can hit him with a riverso to the head, or, if you prefer, a thrust to the chest. But be careful that he doesn't put his left hand to your dagger arm, too. Once you've done this, take four or five steps back and arrange yourself as I told you above.
>
> (Marozzo, 2018, p. 120–121)

Although both opponents are armed and committed to a dagger fight, the operative principles here are similar to those of medieval unarmed dagger defense. Marozzo notes the paramount importance of keeping track of the opponent's dagger hand at all times. The opening attack he recommends is to the dagger hand, and the section includes a grapple of the dagger hand that can finish with a cut to the head or thrust to the chest, with a reminder to beware the opponent's left hand (which will, presumably, be desperately attempting to seize the fencer's own dagger hand). All of this is familiar.

The actual techniques Marozzo includes—a thrust to the weapon hand that wheels into a quick cut—are straight out of Bolognese one-handed sword work. In actual practice, the greater speed of a dagger compared to a sword make a Bolognese dagger duel a very different beast, in which fencers barely have time to snip at their opponents' arms with cuts or light thrusts if they wish to avoid being cut or stabbed in return. Unless the fencer can secure the opponent's dagger arm to open the way for stabs to the chest, such a fight is a (literally) bloody fight of attrition. This is the reason for Marozzo's advice to retreat after delivering a wound, even if it is not fatal. As he explains in his sword and dagger section:

> In those places wherein I have you take three or four steps back, I want you to know that I do so because when you deliver a wound to your enemy, in his fury he may endure having you giving him another of them just to be able to give one to you. But if you take these aforementioned steps back, his rage will diminish, and he will not come at you so brutishly.
>
> (Marozzo, 2018, p. 119)

The dagger is thus a paradoxical weapon in the medieval armory. On the one hand, it was widely regarded as extremely dangerous. Fiore calls it "perilous"; Manciolino's schema makes it the most dangerous of all weapons due to its speed. At the same time, its very speed made it a weapon no sane man would want to fight with, to the point that a challenged party in a duel could almost dare his challenger to meet him with daggers alone in a fight that would almost surely end in both men bleeding from multiple wounds. A medieval dagger fight could be a dizzying display of joint locks and throws, but it could also be a gruesome ambush against which the only real defense was to strike first, as treacherously as possible—or, even better, to not associate with the sort of people who might draw a dagger at the end of a conversation.

For all these reasons, the dagger does not seem to have been regarded as much of a "self-defense" weapon. It was not a tool to preserve one's life, but to enhance one's deadliness: as Fiore says, full of "malice and art."

Sidebar: Which would win: dagger or sword?

In many fantasy stories, there is a character who fights with one or more daggers. This may be a sneaky ne'er do well or even a literal criminal, but just as often, the dagger fighter is a lithe acrobat who can dodge and spin through the battlefield, cutting opponents to ribbons. How much reality is there to this image? Can the dagger truly be a match for a sword?

It will probably not surprise you (although it may be somewhat disappointing) to hear that the answer is no.

Fast as the dagger is, a dagger fencer still has to get within range to employ it. The sword's greater reach allows it to attack first, so a dagger fencer who simply charges a sword-armed opponent must first survive the opponent's attack. In principle, this is a simple matter of using the dagger to parry the sword's cut or thrust, after which the dagger fencer will be upon the sword wielder in the blink of an eye and stabbing away like a homicidal sewing machine. (I have on more than one occasion been asked whether a sword could simply cut straight through a dagger blade. This is utter nonsense; the dagger is still made of steel).

In actual practice, it is not easy to reliably catch a sword blade with a weapon as small as a dagger. The dagger fencer runs a real risk of misjudging the parry and being struck in the hand or arm, which would, naturally, abort the dagger attack. By now it should also be clear that swords do not necessarily follow clear, predictable arcs in combat. Medieval fencing is full of quick redirections of the sword. Even if the dagger does manage to parry the first attack, once the opponent's own movement is factored into the equation (the sword-armed fencer is not, of course, likely to stand conveniently in place and commit everything to a single strike with no associated footwork), it is far from a given that the dagger attacker will be able to close to dagger range before the sword makes a second attack, or even a third.

Armizare discusses this disparate matchup at some length. In all of the scenarios in which the dagger fencer wins, the key to victory is not simply getting close to the sword-armed opponent, but grappling. By now this should be no surprise; grappling is the key to much of medieval dagger fencing. It is important to remember, though, that there is a difference between the correct approach and one that is likely to work. In actual practice, most of the time, the dagger fencer attempts to parry a sword blow and is struck, or successfully parries one attack and is struck by another before fully closing the distance or securing the grapple. It is also worth remembering that a sword can itself be used as a grappling aid (recall Chapter 3), and even an arm encumbered by a buckler can lock an arm (as we saw in Chapter 4) or land a stunning blow that slows the dagger attacker enough to be laid low by the sword.

In short, an opponent with drawn sword is no fit target for an attacker armed only with a dagger. Recall Godinho's rule "against treachery," in which one of the actions the person initiating the dagger attack can perform is to seize the victim's sword hilt before drawing a dagger.

Woe betide the attacker if his victim should see the attack coming in time to back away and draw his sword.

This is not, of course, to suggest that the dagger fencer's case is hopeless. On the contrary, facing a dagger-armed opponent with a sword can be a harrowing experience precisely because the dagger fencer is so likely to act desperately. The balance of advantages is firmly on the side of the sword, but a sword-armed fencer must still be alert and wary against a dagger. Godinho has a personal reminiscence of a man with a *montante*, or greatsword, writing, "I have seen an opponent with sword and dagger, after having his sword broken, gain the *montante* with only the dagger" (Godinho, 2016, p. 99). Even a fencer armed with the largest of swords can be overcome by the "perilous dagger."

CHAPTER EIGHT

Many obligations: greatswords

> *There are many obligations a person must keep when handling and battling with a* montante. *We see in battles that many come to lose by lacking these methods.*
> —Domingo Luis Godinho (Godinho, 2016, p. 99)

In Chapter 6, we discussed fencing with two swords, a staple of medieval fantasy that can only barely be related to the historical Middle Ages. In this chapter, we will discuss another staple of medieval fantasy: the greatsword.

As used in this chapter, a "greatsword" is a two-handed sword that is too large to wear on the belt. Such weapons are roughly the height of a man, or at least the height of the chin. They are generally about 5 to 6 pounds in weight (with some exceptionally heavy examples weighing as much as 7 pounds). A greatsword's blade is generally 40" or longer, while the handle typically has room for at least four hands (about 12" or so), and is not infrequently in the 16" range or longer. Greatswords like this are often associated in the modern imagination with German *landsknecht* mercenaries, and so are often referred to with the German terms *zweihander* or *bidenhander*. Historical fencers tend to use the Spanish term, *montante*, or the Italian *spadone*, since the bulk of extant material on their use comes from Iberian or Italian authors. In this chapter, I will use the word greatsword to describe them all. While there are small differences in physical construction among different kinds of greatsword—blade shape, crossguard length, the presence or absence of parrying hooks, and the presence or absence of side rings—little to nothing of what we are going to say in this chapter depends upon such details. For a discussion of the variation in such protective furniture, see the sidebar below.

Before we go any further, I should make one thing clear: the greatsword (as I am using that term) is not a medieval weapon. This is not the first time we have stretched the definition of "medieval"; Chapter 6 did so as well. But unlike the cloak and dagger, which existed in the Middle Ages even if they were rarely used as companion weapons to a sword, and unlike two swords, the use of which can (barely) be placed in the 15th century through the writings of de la Torre, greatswords truly belong to the 16th century. Nevertheless, giant two-handed swords feature prominently enough in medieval-themed fantasy that I feel this book would be incomplete without discussing them. The greatsword also represents a useful boundary marker when discussing the use of a single weapon against multiple opponents, as it was the Renaissance crowd control weapon *par excellence.*

Although there are multiple Renaissance Iberian and Italian treatises discussing the use of the *montante* or *spadone* at length, in an effort to ground our discussion in the Middle Ages as much as possible, this chapter will focus on Godinho's rules for the *montante* and fencing with what the Bolognese call the "two-handed sword." While these are 16th-century weapons that likely did not exist in the 15th-century version of either escrima comun or Bolognese, this approach can at least ground us firmly in medieval fencing tradition.

> ### Sidebar: The Bolognese two-handed sword
>
> Marozzo, Manciolino, and the Anonimo Bolognese all at least mention the "two-handed sword." Manciolino contains a single reference to it, without any techniques. The Anonimo Bolognese includes what appears to be an incomplete set of techniques for it, while Marozzo treats it extensively, "against pole arms, one-on-one or in company, and in every fashion that may occur while employing the two-handed sword" (Marozzo, 2018, p. 206).

Interestingly, the Bolognese term for the "two-handed sword" is just that: *spada da due mane*. None of the three use the term *spadone*, as do later authors such as Francesco Fernando Alfieri, who in 1653 published a fencing treatise that included an extensive section on "*lo spadone*."

While this might be a simple matter of language evolving over time, there are noticeable differences between Alfieri's illustrations and those of Manciolino and Marozzo (the Anonimo Bolognese is unillustrated). Alfieri's illustrated *spadone* is a massive weapon, fully the height of a man, with a long crossguard and no parrying hooks. Marozzo and Manciolino illustrate two-handed swords that come to the chin, with smaller crossguards; Manciolino's illustration includes no parrying hooks, while all but one of Marozzo's illustrations do. Most intriguingly, Alfieri's treatise is focused on the use of the *spadone* against multiple opponents in a variety of scenarios, much like Godinho's; the Bolognese, on the other hand, focus on the use of the two-handed sword against a single opponent.

All of this suggests that the Bolognese two-handed sword does not fit neatly into the conventional modern categories of greatsword and longsword. It is too big to be a longsword (in the sense of a wearable, two-handed sidearm), but it is too small to be a greatsword (in the sense of a massive sword optimized for use against multiple opponents). From the modern point of view, the two-handed sword straddles both categories. Modern practitioners can and do apply Bolognese two-handed sword teachings to longswords (Marozzo's extended *assalti* for the two-handed sword suggest that he may be adapting traditional forms originally designed for earlier, smaller two-handed "longswords" to a larger, more modern weapon), and they can be applied to most greatswords as well. From the Bolognese point of view, of course, the two-handed sword was what it was, regardless of whether it would fit into our mental categories 600 years later.

Thus, what I say in this chapter about Bolognese two-handed sword fencing could just as easily have been placed in Chapter 3, but it had to go *somewhere*, and placing it here permits us to discuss one-on-one combat with large swords in more detail than is possible when discussing Godinho alone.

The majority of Godinho's discussion of the *montante* is dedicated, like his two sword section, to fencing against multiple opponents. He has only one comment to make about fencing against another greatsword:

> When one has a battle against another *montante*, one should not make use of rules, but of established battle (as two swords), being defended and injuring with thrusts. In no case are *tajos* or *reveses* played (which is counter to many opinions), because in doing them, there will not fail to be great trouble ... keeping in all to the general rule of the first chapter of single sword.
>
> (Godinho, 2016, p. 99–100)

As Godinho says, it is "counter to many opinions" not to cut in a one-on-one fight between greatswords. Monte is one whose opinion on fencing with the two-handed sword differs from Godinho's, basing his two-handed sword fencing on ascending cuts from either side.

Bolognese are not so chary as Godinho of throwing cuts with the two-handed sword either, though they are clearly cognizant of the dangers Godinho discusses. Marozzo does use cuts in his two-handed sword fencing, but he strongly prefers to make initial attacks with thrusts, using cuts only as a follow-up should the initial thrust fail. The seventh part of his first *assalto* provides a good example. The play begins with both fencers standing with their right legs forward and their swords held with their points forward, somewhat on their left sides:

> [S]tep toward his right side with your left leg, and in that step throw a punta incrosata from the outside of his sword, on the right side, which you'll aim powerfully at his left temple. Out of fear of that thrust he'll extend his arms in order to be able to beat it toward his right side with his false edge, and then you, upon seeing this extension, will throw a rising falso dritto to his hands, stepping toward his left side with your right leg as you throw that falso. Without stopping after that falso, throw a tramazzone, stepping toward the enemy's right with your left leg.
>
> (Marozzo, 2018, p. 209)

In this play, Marozzo begins the attack with a thrust from the fencer's left to the opponent's left temple, across the body. The opponent parries with the back edge, but must extend the hands in order to do so. The attacking fencer follows the failed thrust with a rising diagonal cut from the right side to the defender's hands, followed by a "tramazzone" also from the right.

The tramazzone Marozzo refers to normally refers to a wrist cut, but with a two-handed sword it has a somewhat more technical meaning. A tramazzone with a two-handed sword is performed by making a push–pull, crank-like motion with the right and left hands, taking advantage of the leverage provided by the weapon's long handle to quickly whip it around in a wheeling motion. This kind of hand manipulation is less necessary with a longsword-sized two-handed sword, which can be redirected largely with the power of the hips and fingers. The larger weapon, however, can generate significantly more momentum, which the longer handle is an important part of controlling.

Where Marozzo does not initiate an attack with a thrust, he tends not to cut at the opponent but the sword. This is in partial agreement with Godinho's philosophy, for a cut at the sword can be performed at a longer range than a cut to the opponent, and thus the attacker is less at risk of a counter-thrust. The third part of Marozzo's first *assalto* is an example:

> I want you to beat your enemy's sword forcefully toward your right with the false edge ... Having delivered the beat with your false edge, throw a mandritto tondo to his legs, stepping forward with your right foot as you do so, and turn that mandritto into guardia di faccia ... Direct the point of your sword into the enemy's face in such a way that if he were to throw a blow towards your head, you would pretend to parry with your sword, but would let his blow go in vain; and in that same tempo throw a riverso fendente, stepping toward his right side with your left leg as you throw it.
>
> (Marozzo, 2018, p. 215)

In this play, the fencer begins with the sword held low and forward to left, and hits the opponent's sword out of the way with a rising blow. This clears the way for a horizontal forehand cut to the legs. The leg cut would be highly unusual in a longsword fight, for the reasons of reach we discussed in Chapter 3, but with the longer two-handed sword, leg attacks become more practical. The play continues with a thrust at face level, blade parallel to the ground, which is then wheeled into a descending vertical blow as the fencer steps to the left.

Many of the hallmarks of Bolognese one-on-one fencing with the two-handed sword are present here, especially including the long chain of multiple attacks. The attacks in this play are not continuously circular, but they do flow together in a way that is more characteristic of one-handed swordsmanship than the staccato work of a longsword. This is a critical component of fencing with very large swords. A 5- to 6-pound sword might not seem like much weight to control with two hands, but when flung toward an opponent at the end of extended arms as it should be in a proper cut, a greatsword can build up so much momentum that attempting to use it like a longsword can injure the back. Its momentum can be redirected (a critical factor in Godinho's rules against multiple opponents), but it must be carefully guided.

The major exception to this rule is at the half-sword (meaning, in Bolognese terminology, when the swords are crossed). In this circumstance, naturally, the momentum of both swords has dissipated, and both fencers are free to initiate new motions. Perhaps because of the options this opens up, both Marozzo and the Anonimo Bolognese contain extensive lists of *strette*, or techniques at the half-sword, for the two-handed sword (in fact, the Anonimo's two-handed sword material deals exclusively with the half-sword). These strette can be divided into four categories: pommel strikes, grapples, disarms, and blade strikes.

The Anonimo's fifteenth *stretta* of the half-sword, false edge to false edge (i.e., with the opponent's sword to the right of the fencer's sword) provides a good example of a blade action, as well as the longer range at which two-handed sword half-sword play can occur than with longswords:

> [W]hen you are false edge to false edge with your enemy and you have your right foot forward. Feint a mandritto to his left temple. As he defends himself, immediately pass toward his right side with your left foot and feint a roverso to his head, and in this action take your left hand from the pommel of your sword and grip your blade about a *palmo* from the quillons, and strongly drive your sword into that of the enemy, knock it toward his left side, and immediately drive a thrust to his chest.
>
> Here is the counter to the previous attack. When ... your enemy feints a mandritto to your head, pass back with your right foot, throwing a mezzo mandritto to his hands.
>
> (Anonimo, 2020, p. 199)

In this play, the fencer takes advantage of the fact that the opponent's sword is on the right side, and thus the fencer has a clear line of attack to cut the opponent's left temple. The opponent's parry is unspecified, but perhaps the most obvious is to simply push the attacker's cut to the defender's right. Understanding this, the fencer merely feints the initial cut, then steps to the opponent's right and feints another cut, this time to the right side of the opponent's head. This feinted cut forces the opponent to parry again, leaving the sword extended. The fencer punishes this by seizing the sword above the crossguard (between the crossguard and the parrying hooks). Spreading the hands in this way increases the fencer's leverage enough to shove the opponent's sword down and to the attacker's right, opening the way for a strong thrust to the chest.

The prescribed counter is to simply withdraw from the half-sword, striking at the attacker's hands. The fact that the defender can step back to a range at which a regular cut to the hands is possible, along with the double feint of the initial technique, is a good indication that the technique occurs at quite long range despite the fact that it begins with both swords in contact with each other. A natural consequence of the great reach of a two-handed sword is that the swords can be crossed while the fencers are still quite far apart. The Anonimo contains a large variety of half-sword actions that end with a cut or thrust, many of which take advantage of this distance.

This is not to say that all half-sword actions with the two-handed sword occur at such long range, however. Marozzo's first *stretta* with true edge to true edge (i.e., with the opponent's sword to the fencer's left) is a good example, which closes strongly to disarm the opponent:

> If you've come to the half-sword with your enemy, true edge to true edge, send your left hand forward, near the small hilt of your sword, and grab both swords together with your left hand. Thrust your right hand toward your enemy, that is, straight toward the handle of his sword, and grab it with the fingers of your right hand while holding strongly onto the handle of your own sword with your right thumb. Then squeeze both handles together with your right hand while gripping tightly with your left hand above, so that you crush your enemy's right hand and he has to let go of his sword in consequence thereof.
>
> (Marozzo, 2018, p. 226)

The fencer seizes both swords in the left hand, which creates a pivot point so that the fencer can squeeze the opponent's right hand painfully between the handles of the two swords. This action may seem dangerous to the left hand, but can be performed even without gloves. Swords were very sharp, but most of their sharpness came from highly polished cutting faces rather than having cutting faces ground thin like a razor. Because of this, even a blade that is sharp enough to shave with may be seized with a bare hand, and will not cut the hand unless it slides along the edge. Thus, counterintuitively, it is actually safer to press the two sword blades together strongly than to have a looser grip, which might allow the blade to move and cut the left hand. If both hands have parrying hooks (or the "small hilt," as Marozzo says), seizing the swords behind them increases the safety of the action, as that portion of many greatswords was left not only unsharpened but completely dull.

Sidebar: Loaded with options

There are three major defensive features that some greatswords have, but not all: parrying hooks, side rings, and crossguard length. We will briefly tackle these in reverse order.

Medieval swords in the High and Late Middle Ages tend to have crossguards that are as long as their handles. This is a good length for defending the forearms when the sword is held in a point-forward guard, in which the hand and forearm are necessarily at risk.

Greatsword handles are often very long, to give the fencer the leverage necessary to control such a heavy weapon with sword-like speed. This can result in equally long crossguards. Because the hands are spread far apart on a greatsword handle, a very long crossguard is necessary to keep the back hand and forearm behind steel from the opponent's point of view. Despite this, many greatswords have crossguards that are shorter than their handle length. At a certain point, a crossguard becomes so long that some fencers seem to have viewed it as more of a hindrance than a help, relying on other skills to protect their back hands.

Side rings are another defensive feature seen on some greatswords but not others. Side rings are loops of metal that project at right angles from the crossguard so that they protect the fencer's thumb and back of the hand. While the crossguard protects the hands from strikes that slide down the edge of the sword, side rings protect the hand from strikes that slide down the flat. The ring design allows the fencer to place the dominant thumb on the blade to accommodate schools that value such a grip, and also reduce weight compared to a solid or pierced metal disk.

Of course, while side rings do make it harder to "snipe" an opponent's forward hand with quick cuts, they also add weight to a weapon that is already among the heaviest hand weapons.

The last defensive feature we will consider is a pair of small projections partway up the blade, almost like a secondary crossguard. These "parrying hooks" (from the German term, *parierhaken*) or "parrying lugs" can look like vicious secondary blades, but their purpose is primarily defensive. Parrying hooks force an opposing weapon to bind further from the greatsword fencer than the crossguard. This can be a particular advantage against long weapons. Imagine a greatsword with parrying hooks held with the point forward, crossguard and parrying hooks parallel to the ground. A spear that slid down the edge of that blade would contact the crossguard, but the low angle at which a 6-foot sword and a 9-foot spear cross could well allow the spear fencer to stab the greatsword fencer *despite* contacting the crossguard. Parrying hooks force the spear to cross defensive furniture further from the greatsword fencer, increasing the angle of the cross. Parrying hooks can also be used to push against or otherwise manipulate opposing weapons.

Montante are somewhat notorious among historical fencers for having short (longsword-length) crossguards, no side rings, and no parrying hooks. This makes them a useful base case for greatsword fencing, as they represent the stripped-down version of the greatsword. Anything that a *montante* can do, a greatsword with more features can do too, but the reverse is not necessarily true.

Marozzo's third and fourth *strette* form a pair that illustrate other ways of using the two-handed sword at close range. The third stretta is a grappling attack:

> Being, again, true edge to true edge with your enemy, step toward his right side with your left leg, and as you do so, feint a riverso tondo to his head. In this feint, toss your sword behind your back and grab your enemy's right leg with your right arm, meaning that you'll stick your arm between his legs, and put your head under his right armpit so that you'll throw him over your shoulders, and he'll land headfirst.

> (Marozzo, 2018, p. 226)

The fourth stretta counters:

> Whenever someone wants to do some presa to your leg, be aware that any time he throws away his sword to lean over and grab your forward leg, you should quickly draw that leg a big step behind your other one, and throw a fendente at him, or hit him in his back with your pommel.

(Marozzo, 2018, p. 227)

In this play, the attacking fencer steps forward and to the left while pushing the sword to the left, as if preparing a mighty backhand cut to the opponent's face. Since the swords were crossed, this pushes the opponent's sword to the opponent's right. With the opponent's arms thus extended, the attacker flings the sword away entirely and seizes the opponent's extended arm with the left hand. Reaching the right arm between the opponent's legs, the attacker shoots beneath the opponent's extended arm and stands up in a high crotch carry-type position before slamming the opponent backward into the ground.

While it may seem strange to throw away a perfectly good sword in the middle of a sword fight (and one with greatswords, no less), even greatsword fights can end at so close a range that a throw makes good tactical sense. In fact, this throw does not work without discarding the sword entirely, for the attacker's left hand needs to extend the opponent's right arm so the attacker can shoot under the armpit. It is also worth noting that this throw, like many of Marozzo's throws, is designed to throw the opponent "headfirst" to the ground—a dangerous technique.

At the same time, the fourth stretta points out two risks to such a technique. The first is that the attacker shoots forward for the leg grab at too great a range, allowing the opponent to step back and deliver a downward cut to the now-defenseless would-be grappler. In the heat of a fight, such a miscalculation is easy to make. If the range is appropriate for a grapple, the defender can still strike the attacker in the spine with the pommel.

One last example of half-sword work with the two-handed sword will bring our discussion of one-on-one fencing with the greatsword to an end. Marozzo's 12th stretta of true edge to true edge is one of my favorite Bolognese techniques:

> Once again being true edge to true edge with your enemy, step well forward with your left leg … In this step, swiftly kick him in the testicles with your right foot … and thus you'll have accomplished two effects: namely, you'll have given him a kick, and also performed a gallant presa.
> (Marozzo, 2018, p. 229)

This technique demonstrates several elements of greatsword fencing, as well as illustrating how the historical definition of "gallantry" could include techniques that to modern eyes seem "dirty." The initial step forward and to the left is important for more reasons than because it sets up the kick to the testicles; it also pushes the opponent's sword back and to the right, maintaining control of it. Pushing obliquely rather than straight back ensures that the fencer is not pushing against the opponent's full strength. This is arguably the most important element of the technique; with the opponent's structure thus compromised, the fencer could just as easily throw a backhanded cut at the opponent's neck. The kick to the testicles is an opportunistic way to capitalize on the body position created by collapsing the opponent's weapon backward and to the side.

Let us now leave one-on-one fencing with the greatsword behind, and turn to fencing with it against multiple opponents. It is in this role that the greatsword truly shines. Few other historical weapons combine the greatsword's ability to threaten a large area with the ability to redirect its attacks as quickly as a sword.

Both elements are critical to the greatsword's role as a crowd control weapon. Unlike the cuts of two swords, which can surround a fencer with a rapid but short range flurry of blows, the greatsword's cuts are comparatively ponderous (I am speaking in relative terms: in absolute terms, the cuts of a greatsword can still come eye-wateringly fast). However, the great reach of a greatsword means that it can keep opponents very far at bay. A single fencer armed with a greatsword can hold off a great many opponents armed with shorter weapons.

A spear or other polearm could do the same, of course. However, polearms are not generally balanced such that they can be easily redirected if swung in a wide arc (a cutting polearm can be reversed quite quickly, but only if the hands are spread far apart on the haft, robbing it of reach). The greatsword, for all its size and weight, is still balanced proportionately like a sword. This makes it significantly handier when swung in the large sweeping cuts that are necessary to keep multiple opponents at bay, and thus more able to quickly redirect and strike at an attacker who is attempting to sneak into range of the greatsword fencer.

As with all multiple-opponent scenarios, crowd psychology plays a significant role here. Utterly fearless opponents could make short work of a greatsword fencer while suffering only one or two casualties. Real people, however—and especially real people in a street fight—are not likely to be utterly fearless. On the contrary, even in sparring scenarios where the swords are blunt and the fencers are wearing protection, there is a powerful incentive to let someone else be the first to come into reach of an angry, active greatswordsman. It this hesitation that a greatsword fencer largely plays upon when holding off multiple opponents.

While a greatsword has many advantages compared to smaller swords, actually utilizing those advantages is no simple task. It is still a single weapon, albeit a large one. Successful greatsword use against multiple opponents requires strict economy of motion as well as the *correct* motion for the situation. A greatsword fencer must always be on guard against waving the sword about in a superficially impressive display without actually threatening the opponents in the necessary manner. Godinho warns against this risk at the beginning of his chapter on the *montante*:

> There are many obligations a person must keep when handling and battling with a *montante*. We see in battles that many come to lose by lacking these methods; neither from dropping it from being badly wounded, nor from perhaps tripping and falling, nor from unfortunate events that a man cannot resist, but by being over-taken by opponents armed with only a sword, and others with a sword and shield.... It was and is the reason why, in place of true rules, they make flourishes, which are necessary only for masters, while the rules for the players are as follows.
>
> (Godinho, 2016, p. 99)

In this passage, Godinho contrasts mere "flourishes" with the true "rules" of the *montante*. He is warning that all of a *montante*'s motions in actual combat must be strictly utilitarian. Anything less gives opponents an opening that the greatsword fencer cannot afford to give. He goes on:

> For he that will learn the *montante* in order to defend his person and injure his enemy or enemies, does its rules at the time of battle with much sense and caution, making each one in its necessary place (which is why they are culled "rules"); for instance, he does not do the withdrawing rule when surrounded, nor the surrounded rule when having opponents in front, nor with too much speed, except when pressed.
>
> (Godinho, 2016, p. 99)

In other words, the right move at the wrong time can be just as useless as mere "flourishes."

For the knowledgeable fencer, however, the *montante* was a formidable weapon. Godinho's second rule for the *montante* gives an idea of how versatile this weapon could be in the right hands. He writes:

> When in a narrow street, surrounded on both sides ... put your back against one of those walls. Turning the *montante* at the wrist so that it is above the crook of the left elbow of the left arm, which remains withdrawn to the chest, put the right foot to the left side with much speed, and at the same time extend the arms, giving a nails-up thrust to those on that side. Just as the thrust is given with the foot put in and the back against the other wall, turn the point of the *montante* in the air, bending at the wrist, above the left arm. In

the manner given above, put in the right foot with a nails-up thrust to those on the right side, ending with your back on the other wall.

(Godinho, 2016, p. 100)

In this rule, the fencer is surrounded in a street too narrow to throw any cuts—almost an alley. We might expect this circumstance to be a disaster for someone carrying so large a weapon, but it is not so. Instead, Godinho makes use of the greatsword's length to throw powerful thrusts to one side of the alley and then the other, rotating the sword overhead to maneuver it in the narrow confines of the street and powering the thrusts with rotating steps that turn the body in a half circle counterclockwise. The resulting movement fills the alley with steel, giving opponents no opportunity to close.

Interestingly, Godinho notes that this rule has a maritime application as well:

> The same rule will be used in a gangway of a galley, the slave benches becoming the "walls" that are dealt with in the beginning of the rule, but this is understood as the opponents being in control of the stern and bow.
>
> (Godinho, 2016, p. 100)

This is a good example of the *montante*'s versatility as both a military and a civilian self-defense weapon. It also illustrates how *montante* rules can and should be adapted to analogous contexts. The operative principle of the second rule is not that the fencer is in a literal narrow street, but that there are opponents on either side and the fencer cannot throw any cuts—whether the reason is because there are literal walls that would obstruct the cuts or because doing so would endanger the fencer's own side.

Sidebar: Clearing a house

Escrima comun's rules for two swords and *montante* are a treasure trove of information about practical sword fighting in a variety of urban environments. Somebody once asked me what medieval "house clearing," or fighting room to room inside a building, looked like. While I can think of no medieval (or Renaissance) fencing or military text that discusses this problem directly, Godinho's 13th rule, for fighting at a crossroads, is illuminating:

> If I am told that one necessarily has to turn close to the corner, it is easy to send the *montante* in front, stitched to the wall with a thrust, extending the arms as much as he can, and then continue the rule that will be necessary for the occasion. The rest is folly, making flourishes while a person's life hangs in the balance.
>
> (Godinho, 2016, p. 108)

In its intended application, the rule works like this: in order to turn a street corner to the left, the fencer would send the sword down the left side of the alley or street being turned into, to ensure that no attacker was waiting just behind the corner. It is not hard to imagine this same rule being applied (perhaps with a smaller sword) to the halls of a house or turning a corner into a room, where an attacker might be waiting just inside the door. Likewise, the second rule, on fighting in a very narrow street, could also easily apply to fighting in a hallway.

Questions like this are often anachronistic in how they imagine medieval warfare, but they are not just interesting fantasies. They are also a good example of how escrima comun rules can be extrapolated to scenarios other than the ones for which they were written.

Another scenario in which a greatsword fencer had to be cognizant of wounding an ally was escorting an ally through a hostile crowd. Godinho refers to this circumstance as "guarding a lady," perhaps to indicate the noncombatant status of the person being escorted. The scenario is one that may have been of particular interest to greatsword fencers, as the greatsword's abilities against multiple opponents made it ideal for bodyguards. The rule proceeds as follows:

> Have the woman or friend that you intend to keep safe attached to your belt from behind, advising them to not exert force on it, only so that they remain safely behind your back. Then play a *tajo*, putting in the right foot, and a *reves* putting in the left foot, which do not leave flying toward the rear, so as not to injure the one that is guarded; it will only come to the shoulders when it is armed ...
>
> If they will surround him, he makes use of Rule Seven ahead, sitting the person down.
>
> (Godinho, 2016, p. 103–104)

The basic principle of the rule is straightforward. The escorted lady (or "friend") is placed at the fencer's back, and the greatsword sweeps a cone of space ahead of the pair (Godinho gives explanatory notes on how the bodyguard should use cuts or thrusts depending on how many opponents are present). This rule provides insight into an eminently practical scenario, and also gives an idea of the severe constraints a greatsword fencer might need to operate under. As Godinho notes, it is only practical to move the escorted person when all the enemies are in front; if surrounded, the escorted lady (or "friend") must drop to the floor so the escort can address threats from all corners. Otherwise, there is no way to address threats to the rear without undue risk of hitting the escorted person. Attempting to turn both escort and guard together is a recipe for disaster (which I can confirm from sparring experience).

This might seem counterintuitive: if the escort needs to turn, why can't the escorted person simply turn as well? The answer is that, while *montante* rules may seem regular and easily anticipated on paper, in practice there are myriad subtleties of speed—small delays and bursts of motion—that keep opponents on their toes, to say nothing of the fact that a skilled fencer may need to switch rules without warning. The same unpredictability that makes a greatsword an effective deterrent against multiple opponents also makes it very difficult for the escorted person to synchronize with if he or she is not safely attached to the fencer's back with a hand on the belt, and that handhold limits the fencer's ability to maneuver. Thus, in practice, if the escorted person cannot join the fight, attempting to synchronize with the escort's maneuvers results in either being struck by the escort or creating an opening in the escort's defensive pattern with fairly high frequency.

This rule is also particularly interesting for the way it varies among attacks. The escorting fencer must balance two concerns: covering a wide enough arc to deter all potential attackers, on the one hand, and keeping the attacks coming in succession as quickly as possible to minimize the opponents' windows of opportunity, on the other hand. The rule is structured so that the escorting fencer throws wide cuts (coming to the shoulders, covering a nearly 180-degree arc) if opposed by many attackers, narrow, quick cuts (thrown with the wrist) if opposed by only two attackers, and thrusts if opposed by only one.

As we saw early in the chapter, Godinho strongly advocates against using thrusts with the *montante* for multiple opponents, and strongly advocates against using cuts for single opponents. His tenth rule, for fencing when surrounded in a plaza, field, or street, is another good example of this rule and of a circumstance in which he breaks it. It also contains a historical example of something you might have thought was a mere videogame convention: a spin move.

> One must have much acuity, speed, and great vigor when surrounded by opponents in a battle. One quickly takes up his *montante* … in the middle of the attackers, he cuts a *tajo*, bending at the knees with the head straight, established on the left foot. Then he cuts another, putting in the right foot, in a manner that he goes sideways, alternating left and right foot, cutting *tajos* in a continuous motion, circling himself in a wheel with the *montante*, giving a *tajo* at each step, which may be up to three or four, and not passing five because of the danger that it can have in weakening the head. These steps with *tajos* finished, he returns to where he started with other steps with *reveses* …
>
> Take note that if you will be surrounded in a wide field, that when you give the steps, they aren't always to one side, returning to where you had left, but continually stepping first to one side, then to another, running the whole circuit in a wheel.
>
> (Godinho, 2016, p. 106)

In this rule, the fencer makes forehand cuts while turning counterclockwise and backhand cuts while turning clockwise, in each case making continuous pivots (Godinho is keenly aware of the risk of dizziness here, which is not only why he advocates making no more than five cuts without reversing the direction of rotation but also why he specifies that the knees may be bent but the head should remain "straight").

It may seem fanciful to spin about a battlefield with a giant sword, but all the pieces of Godinho's rule have a purpose in the scenario of a fencer surrounded in an open space. The steps in this rule travel even as they cause the fencer to spin 180 degrees with every step. This travel is important to allow the fencer to exert pressure on the cordon of attackers. It is also an important component in allowing the fencer to make and exploit space. Godinho is careful to specify that the fencer's cuts should "circl[e] himself in a wheel," which is important to ensure that the fencer is at least operating within a circle of space whose radius is equal to the length of the extended arms and sword. However, a greatsword fencer who was surrounded by opponents kept at bay no further than the length of the sword—about 5 to 6 feet—would quickly be overwhelmed. At that distance, even fencers with shorter weapons are in range to strike with a single step. To keep opponents further at bay, the greatsword fencer must move. To understand how, we should break down the mechanics of one of Godinho's Rule Ten spins.

The spin begins with a forehand cut powered by the fencer stepping forward with the right foot. This is, so far, an ordinary forehand cut, and should either strike an opponent ahead of the fencer or cause that opponent to spring back in fear of the cut. It is also an opportunity for an opponent to the fencer's rear to close in and attack. For this reason, Godinho does not have the initial cut travel only through a forward arc (as he does, for instance, when guarding a lady). Rather, the fencer continues to pivot counterclockwise on the right foot, causing the forehand cut to sweep behind the fencer, thus striking or deterring any attackers approaching from the rear. With the rear attacker(s) thus temporarily halted, the fencer takes another step forward with the *left* foot as the spin continues, pressing forward and leaving the rear attacker(s) in the dust. Just as the attackers have figured out the fencer's pattern of movement, the greatsword fencer charges off in another direction with backhand cuts and clockwise spins.

When executed with "much acuity, speed, and great vigor," as Godinho says, this rule can effectively isolate a section of the attackers' cordon to press upon and leave the attackers constantly scrambling in the fencer's wake. At the critical moment, the pattern changes yet again:

> If it happens that you have to break the squad of enemies that keeps you surrounded, press with the step to the side where you see more weakness, and arriving close, shout with a loud voice, and at the same time, give a nails-up thrust to that side with a big leap in a circle, and at the end of the thrust, a *tajo* and *reves*.
>
> (Godinho, 2016, p. 106)

In the original, Godinho specifies shouting "*Afuera!*" or "Get out!". When the time comes to break the opponents' cordon, the fencer finally attacks with a thrust, skewering or pushing back a particular opponent, then leaping into the space thus created. The circle of enemies now broken, the fencer can withdraw.

Sidebar: Beating pikes

If you are at all familiar with greatswords, you have probably heard of them being used by Renaissance soldiers to clear paths for friendly troops to attack enemy pikemen by beating aside enemy pikes. When greatswords come up in discussions with the public, this topic invariably arises. Sometimes it seems the only use for greatswords that people can imagine is sweeping aside an enemy's pikes to allow your own pikemen free rein to charge home. People are often surprised to hear that fencing treatises—even those that clearly contemplate the use of the greatsword on the battlefield, as Godinho's does—do not discuss this usage at all.

From the standpoint of practical experiment, I admit to a healthy dose of skepticism about the whole affair. A greatswordsman whose job is to sweep pikes out of the way occupies more frontage in a battle line than does a pikeman, with the inevitable consequence that even a line of greatswordsmen advancing before friendly pikes will see each greatswordsman targeted by multiple enemy pikemen. And while a greatsword's long blade certainly can contact multiple pikes, it is a mistake to imagine pike formations as staid, static affairs in which the pikes are simply leveled at the enemy and held at the ready without any active use at all. It is difficult enough to catch one pike—provided it is being handled by an active, aggressive pikeman who knows how to operate his weapon—with a greatsword, let alone several. And even if successful in disrupting multiple pikes, a greatsword can only disrupt the front rank of pikes. Subsequent ranks of pikes still project beyond the front rank of men.

I do not wish to be understood as saying that greatswords *cannot* beat aside a row of pikes, or never did. However, it is important to understand that they have significant limitations in this role. Perhaps for this reason, their principal use in our extant fencing treatises is not against polearms (though see Chapter 11 for a discussion of greatsword vs. heavy polearms). Rather, the greatsword is presented as a powerful weapon for single combat and for occupying space against multiple opponents.

At this point, you may be wondering what happens if one of the attackers interrupts a rule's flow by simply parrying the greatsword. Wouldn't this give other attackers a window of opportunity to attack?

There are several possible answers to this dilemma. The first is that, because a greatsword typically outranges other weapons in a multiple-opponent scenario, a greatsword fencer is usually free to change the trajectory of an attack to avoid a parry or make it very difficult. For instance, when Godinho calls for a *tajo* or a *reves*, he does not usually require that these cuts be thrown along a particular trajectory toward a particular target. A cut that was intended for the head could often (granted, not always) just as well be thrown to the legs, or vice versa, to avoid a parry. The second answer is that greatsword cuts land with thundering force. They certainly can be parried (even by a dagger, if the parry is done well), but the parrying fencer must be sure to catch the greatsword at the base of the parrying weapon with good form or the parry will have insufficient leverage to stop the cut. Parrying a greatsword is a daunting task, one which is subject to the same "you first" crowd psychology that the outnumbered fencer is depending upon. The third answer is that, because parrying a greatsword is so demanding, the outnumbered fencer can usually depend upon the true attack to come from an opponent other than the one who stepped into harm's way to parry the whirling sword, and react accordingly.

But the fourth and final answer is that yes, an opponent who is bold enough to parry a greatsword mid-rule places the outnumbered fencer in great peril. Formidable as it is, not even a greatsword can completely mitigate the risk inherent in fighting multiple opponents. It is a powerful but precarious weapon, unusually demanding on a fencer's physical fitness, form, and tactical acumen.

Which would win: Greatsword vs. sword and shield?

The choice of greatsword and sword and shield seem, at first glance, to be opposite weapon combinations. The one is an enormous, aggressive-looking sword; the other seems solid and imperturbable. What would happen when one of the most unstoppable swords clashed with one of the most immovable defenses?

The reality, of course, is that both of these characterizations are misleading. In many ways, the greatsword is an extremely defensive weapon, one whose true forte is denying space to opponents. The sword and shield, by contrast, can be extremely aggressive by permitting a fencer to charge into the opponent's reach while under cover. Yet this only makes the matchup more interesting: if both weapon combinations lend themselves to aggressive or defensive fencing depending upon circumstance, which dynamic would prevail when the two are pitted against each other?

Evidently Godinho considered the question interesting as well, for he devotes an unusual amount of attention to the contest of *montante* vs. sword and shield. In the second chapter of his sword and shield material, he warns the sword and shield fencer:

> Take note that if one is attacked by a *montante*, that it is possible to battle safely, but being able to withdraw is best; when one does not, take the parry on the shield and attack below with a thrust.
>
> (Godinho, 2016, p. 55)

In light of this advice, Godinho's fourth rule for the *montante*, "Against a Shieldman," makes particular sense:

> [P]utting in the right foot, play a *tajo* to the legs of the shieldman. The *tajo* cuts until it turns above your head, and the *montante* ends armed for a *reves*, nails-up with the point to the floor, the quillons covering your head. Then, putting in the left foot, play a *reves* to the legs, giving a turn with the *montante* until it stops above your head with the point to the floor and the quillons covering your head. The point that ends at the floor like this has to be nails-down, because then the *tajo* is armed. ...
>
> Note that the points that I make mention of in this rule are not committed thrusts, just armed, so that the shieldman does not slip in below.
>
> (Godinho, 2016, p. 101–102)

In this rule, the greatsword fencer throws a descending diagonal cut from the right, allowing the cut's momentum to continue along its trajectory until the sword is in the fencer's upper right, then guiding it from right to left so that the swords end high and on the fencer's left side with the point forward and down. The rule then cuts from left to right, swinging around to the left again and ending high and on the right with the point forward.

Godinho thus addresses both the major weakness of and the obvious strategy for a fencer armed with sword and shield. By cutting to the legs, the greatsword fencer can largely bypass the opponent's shield. As we saw in Chapter 5, most shields have a very difficult time defending the lower legs, and even those that reach that far down may have difficulty stopping the formidable momentum of a greatsword cut. Against most swords, this is not a significant issue, as a cut to the legs is necessarily shorter-ranged than a cut to the head. A greatsword's blade, however, is long enough to threaten the legs at a reasonable distance. At the same time, by allowing the greatsword's momentum to carry through and around until the point is once more forward and threatening from the upper right and left, the rule makes it difficult for the sword and shield fencer to do the obvious thing and simply rush in low under cover of the shield. The fact that the rule threatens thrusts from above and from the side, rather than simply from above, makes it particularly challenging for the sword and shield fencer to parry the threatened thrusts without opening the shield so much that the greatsword fencer can simply slip around the shield and stab to the opponent's center.

None of this is impossible for a sword and shield fencer to deal with, of course—as Godinho himself notes, it is *possible* to battle safely against a greatsword. It is simply inadvisable. As the legs are the obvious target and difficult to defend with either sword or shield, the sword and shield fencer must be especially mindful to hover at a distance from the opponent far enough away that cuts to the legs can be defended by slipping the leg, yet close enough to rush into distance once the cut to the leg has passed.

The timing issues inherent in this delicate dance of distance are suggested in rule five, which Godinho gives against two or more shieldmen. The cutting pattern in this rule is the same, but with a half-turn to the left or right at the end of each cut, so that the greatsword sweeps a larger arc and the points can be directed anywhere in the fencer's forward arc. Godinho notes that

> The turn of the blow is very necessary to arm said points, which serve to parry the blows that the opponents are able to give as soon as the *montante* passes, and the posture of the point impedes the entrance below the *montante*; this should be done quickly.
> (Godinho, 2016, p. 103)

In other words, once the greatsword has passed from right to left, the sword and shield-armed opponent may be expected to enter into distance and throw an attack, which the greatsword parries as it travels around the fencer from right to left in order to "arm" the threatened thrust—if, that is, the greatsword fencer is fast enough. It is certainly possible that the sword and shield fencer is fast enough to strike as the greatsword is traveling around to cut to the thrusting position (the "turn" of the blow that Godinho says is "very necessary"), or that the sword and shield fencer can use the shield to impede the greatsword as it "turns," creating an opening for the sword. Neither fencer is entirely free to give in to unbridled aggression, but neither can afford a purely defensive stance that waits for the opponent to make a mistake.

Nevertheless, the initiative lies with the greatsword, and the sword and shield fencer's options all depend upon being able to avoid the leg cut by a slim margin. No wonder, then, that Godinho recommends that the sword and shield fencer withdraw from such a contest if possible.

All of this may make us wonder why anybody ever bothered to fence with sword and shield at all, if the advantages of that weapon combination could be so nullified by a greatsword. Here it is important to remember that the greatsword, as we have been using that term in this chapter, is quite a late development, one that essentially postdates the Middle Ages. As we saw in Chapter 3, a *shorter* two-handed sword—one too short to cut an opponent's legs from a comfortable distance—is at a significant disadvantage against a sword and shield, while a sufficiently long polearm, such as a two-handed axe, is unlikely to have the necessary speed of redirection to prevent the sword and shield fencer from entering (the greatsword's sword-like point of balance is critical to allow the rule to be "done quickly" enough to work). In other words, for most of the Middle Ages, sword and shield fencers fighting two-handed weapons did not face the particular threats that a greatsword presents.

It is true, however, that greatswords coexisted with the sword and shield in the 16th century. To understand why the sword and shield was not simply discarded in favor of greatswords once they came on the scene, it is useful to remember two facts. The first is that a sword and shield lends itself to the synergies of close-order formations much more does a greatsword, and is far more useful against missile weapons. This makes the sword and shield a more generally useful weapon combination for military use than the greatsword. The second is that in order to deal with a sword and shield, the greatsword must prioritize low cuts. While this works, it also reduces the greatsword's effective length … and thus its ability to clear space around the fencer against multiple opponents. The sword and shield thus reduces the utility of the greatsword against multiple opponents more than most weapons, making it an obvious choice for fencers who expect to outnumber a greatsword. As Godinho says, it is best to withdraw when "one" is attacked by a *montante*. Multiple shieldmen are a much more difficult proposition.

CHAPTER NINE

The unseen: what about axes and maces?

Chapters 3, 4, 5, 6, 7, and 8 have all been concerned with some form of sword (or, in the case of Chapter 7, daggers, which are close enough to swords for government work). But swords were far from the only hand weapon used in the Middle Ages. What about weapons that *don't* look like swords, such as axes, maces, and warhammers? What about polearms?

If I were reading this book as a teenager, the omission of these weapons up to this point would surprise me. When I was first learning about medieval weaponry, I assumed that weapons, and especially one-handed weapons, fell into three major categories: swords, axes, and bludgeons (maces or hammers). These were the three types of weapons I saw most commonly depicted in fantasy artwork, in games, and in other media. I spent a good many hours wondering just why a medieval warrior might choose to use a sword, axe, mace, or warhammer. What were each of these weapon types for? I had a vague notion that maces and warhammers were designed to fight armored opponents (but how were they different from each other?) while swords were somehow better at fighting unarmored or lightly armored opponents (because they did more "damage?"). And as for axes ... perhaps axes were something like a compromise? Or maybe axes offered more options for controlling an opponent's weapon or shield, since they could be used to hook and pull.

At the time, it was all a great mystery, and as I took my first steps into the world of historical European martial arts, I looked forward to finding out the *real* answer. You may imagine my surprise when it turned out that our extant treatises contained virtually nothing about how to fence with axes, maces, or warhammers.

Or rather, our extant treatises contain virtually nothing about how to fence with *one-handed* axes, maces, and warhammers. The use of the two-handed axe or warhammer, in the form of the poleaxe, halberd, and similar weapons, was a matter of considerable interest to medieval fencing masters. We'll discuss the use of such weapons in Chapter 11.

But what about the one-handed versions? Do historical treatises really have nothing to say about how to use, say, an axe and shield? What about a mace and shield? Or (my teenaged self would ask hopefully) dual-wielding axe and warhammer?

Alas for my teenaged curiosity, the answer is largely no. One-handed axes appear nowhere in our extant medieval martial arts, and one-handed bludgeons are given only the briefest treatment. A brief treatment is better than nothing, however, so in this chapter we will discuss what little is said about fencing with one-handed bludgeons.

We have already seen the distinctly German tradition of fencing with longshield and club in Chapter 5. Interestingly, KDF appears to make almost no distinction between fencing with longshield and club or mace and longshield and sword. The focus of masters writing about that weapon combination is typically not how to use the "offensive" weapon, but how to use the very unusual shield (including, as we saw, using the shield in two hands as an offensive weapon in its own right). However, Paulus Kal does distinguish between them at one point:

> [Do] thus the first technique behind the shield if you have a sword or mace: if you have a sword strike whatever opening you see; or cast [the mace] and call your marshal to give you another mace.
> (Kal, 2019)

This strongly suggests that Kal did not consider the sword suitable for throwing, whereas the mace was—enough so that the rules of a longshield duel evidently included provision for supplying combatants with a certain number of maces to throw.

Maces may seem like they would be far too heavy to throw. Contrary to the fantasy stereotype of large, heavy maces, though, there is little to no difference in weight between medieval maces and one-handed swords. Maces simply concentrate their mass at the end of their haft, giving their percussive blows more force, while swords distribute their mass more evenly throughout the weapon. By corollary, maces are typically shorter and more compact than swords, which is part of what makes them more suitable for throwing.

Bludgeon-throwing can also be found in armizare. One of Fiore's plays concerns a fencer with two clubs and a belt dagger being attacked by an opponent with a spear. In this unusual circumstance, Fiore advises the club fencer to throw one of his clubs to provide cover as he rushes forward to employ his shorter weapons:

> This Master defends with two cudgels against a spear, as follows: when the spear man approaches to attack, the Master throws the cudgel in his right hand at his opponent's head. Then he quickly strikes with the cudgel in his left hand to make cover against the spear, then he strikes his opponent in the chest with his dagger, as is shown next.
>
> (Fiore, 2017, p. 31v)

Fiore does not elaborate upon the circumstances in which this might occur. However, the fact that both clubs are illustrated as rough-hewn tree branches suggests that the club fencer has been surprised and is using improvised weapons. This in turn suggests that throwing a mace or club was useful in more circumstances than the ritualized context of a longshield duel. Modern sparring experiments—to say nothing of the thriving modern axe-throwing community—suggest that this technique is equally applicable to one-handed axes.

So much for throwing. What about techniques using a bludgeon (or axe) in hand?

Pietro Monte is the only medieval source to discuss one-handed bludgeons in their own right (as opposed to as accompaniment to a longshield). Interestingly, he makes little distinction between maces, clubs, and warhammers. According to Monte, the weapon he has in mind is popularly referred to as a "mace" (*maza*), but its technical name, in Latin, is a "club" (*clava*). However, his description makes it clear that he is speaking of what English speakers would call a "warhammer," with a clawed hammer face, a rear-facing point, and a top spike.

> [I]ts iron should be like a poleaxe or *tripuncta*, one part of the hammer which is larger should be divided into three small points in the way of a diamond. The other part should be like a little lemon, and those points should be fixed in the haft with a knot at the middle, and above or in front a short point should be placed.
>
> (Monte, 2018)

Despite this detailed description, none of the techniques Monte offers *require* the hammer to have a distinctive hammer configuration. The warhammer, according to Monte, is not a subtle weapon. He has only two techniques for it, the first of which is simply to grab the weapon with both hands and hit the enemy as hard as humanly possible:

> [T]aking the club with both hands, three or four blows are to be struck with the utmost force and velocity.
>
> (Monte, 2018)

Although this is a simple technique (in fact, it's hard to call "hit him as hard as you can" a "technique"), the use of two hands on what is ostensibly a one-handed weapon is a fascinating insight into the nature of warhammers. Monte has two reasons for advising the use of two hands on the warhammer. The first is, as he says, that a fencer with one hand on his weapon will have a hard time parrying a fencer who uses two hands—even to the point that he may lose hold of his weapon entirely. His second reason comes in a discussion of how fighting in full "white armor" (i.e., plate armor) changes the fight.

An opponent in full armor, Monte says, can simply ignore all but the strongest blows. Thus, when armored combatants are close to one another, striking weapons become useless. Interestingly, Monte says that this is even true for clubs and axes, which by nature strike with more percussive force than swords. As he says, "we can never apply great blows when he always turns aside or enters in where we can make a small blow on him; which he who is entirely in white armour cares nothing for" (Monte, 2018). Elsewhere, he even recommends parrying the blows of warhammers with an armored forearm, which also suggests the relative impotence of warhammer blows against plate armor: "if he throws any blow with a sword, *estoch*, club, or similar, we should receive it with our sword *or arm* (Monte, 2018, emphasis added). In other words, even a warhammer is little good against plate armor unless a fencer can maintain his distance enough to deliver full, powerful blows. This is a significant limitation on the utility of concussive force in armored combat, as we will discuss in greater detail in Chapter 12. No wonder that Monte advises the use of two hands on the warhammer, to eke out every last ounce of force from its blows.

The second technique Monte gives for the warhammer makes use of the weapon's belt clip to hook an opponent and pull him off balance:

> This club, or *acuscula* (little needle) in the vernacular, should, besides the three points, that is, one with a blunt point, another in the middle, and the third sharply pointed, also in the nail which joins the wood to the iron, should be a very strong hook or crook, and from this hook the mace is suspended, and when there is fighting it can take the adversary by the neck or other limb, and drawing strongly, especially by turning around, as should be done at the same time as the horse, and in this way it is easy to draw the adversary out of the saddle. ... [E]ven a footsoldier, when there is fighting with axes and white armour, can carry a hook in the axe to take the adversary by the neck, since it will be easy to throw him, and then he cannot work any great or dangerous blow against us.
>
> (Monte, 2018)

It is interesting to note that Monte uses the belt hook, rather than the back spike, to hook an opponent. The back spike of a warhammer is not sharply curved, which makes it suboptimal for use as a hook (note, though, that Monte specifically recommends that the belt hook of one's hammer be made strong enough for use as a weapon, which suggests that not all belt hooks were suitable for such a purpose). Monte does not discuss one-handed axes, but modern experience strongly suggests that the "beard" of a hand axe can be used the same way to hook, pull, and imbalance.

These three techniques—throwing the mace, striking with two hands, and hooking the opponent's neck or limbs—constitute the sum total of surviving medieval techniques for one-handed maces and hammers (or "clubs," if we are to use what Monte says is the technical name).

What are we to make of this? Was fencing with bludgeons so straightforward that no additional teachings were necessary? Or is something else at work?

Our available evidence is limited (that's why we're in this situation to begin with), but I tend to think that bludgeon fencing is not a straightforward matter of bash and smash. It's important here not to lose track of the forest for the trees. Recall that our longshield material largely treats the sword and club as interchangeable weapons. In other words, anything you can do with a sword (or, at least, with a sword and longshield in hand), you can do with a club or mace.

This is not surprising. While this book is organized by weapon, as we discussed in Chapter 2, medieval traditions themselves did not organize their teachings in this way. Rather, traditions taught the bulk of their principles with a single weapon or weapon combination, and then taught other weapons with much smaller lessons focused on how

---

Sidebar: What about hand protection?

One seemingly obvious difference between a sword and a bludgeon or axe is that the hilt of a sword protects the hand with a crossguard, while the haft of a bludgeon or axe offers no such protection. It might seem that this would significantly impact the way one would fence with the weapon. After all, how can you parry a sword cut with the haft of a mace just as you would with a sword, if the opponent's sword might slide down to strike your fingers?

There are four answers to this issue. The first, of course, is to admit that a haft that does not protect the hands and *does* add an element of risk to certain actions that would not be present with a sword. When parrying with a guardless haft, a fencer must make greater use of strong, percussive parries—hitting incoming blows away, rather than redirecting them with the haft of his weapon—to keep his hands safe.

The second answer is to realize that, with wooden hafts at least, sharp blades will tend to bite into the wood, which will inhibit their ability to slide down the haft. This is an aspect of historical fencing that is not well represented by our sparring weapons (which are, of course, blunt).

The third answer is to note that axes and bludgeons were less often carried as "civilian" weapons than swords were, and thus a fencer armed with an axe or bludgeon was reasonably likely to be wearing armor of some sort.

But the fourth answer is that bludgeons and axes need not have less hand protection than a sword. A simple disk of metal above the hand (often called a rondel) will prevent an opponent's blows from sliding into the hand, allowing the weapon to be used just like a sword. Monte specifically recommends this in a warhammer, and it is not hard to imagine why.

that weapon differed from the baseline or exemplar weapon. To the extent that we have little material focused on maces, hammers, and axes, we may tentatively conclude that medieval fencing traditions did not think of these weapons as very different from swords in the way that they were used.

We should pause here and clarify that the *way* a weapon is used is not the same as what a weapon is used *for*. Bludgeons and swords wound through different mechanisms (as do axes and swords, to a lesser degree), and a sword's ability to wound with a percussive blow through armor is certainly less than is a bludgeon's or an axe's. Yet all of this can be true without changing the tactical or mechanical fundamentals of cut, thrust, and parry that a fencer would learn with a sword.

There is, however, another possibility as to why we have so little material dealing with one-handed bludgeons and (especially) axes. It could be that these weapons *were* very different from the swords and polearms that is the subject of the majority of our extant material, and we simply do not have any of the material that would provide evidence of that. It could be that texts focused on these weapons have simply been lost to the vicissitudes of history, with only scattered references remaining.

This is not completely implausible. *Le Jeu de la Hache* is an early 15th-century text focused exclusively on the use of the poleaxe, which we will discuss in more detail in Chapter 11. As we saw in Chapter 4, the Walpurgis Fechtbuch is focused exclusively on sword and buckler. The Middle Ages did produce some focused, single-weapon treatises. Perhaps it is simply bad luck that no such treatises focused on one-handed axes or bludgeons has survived.

The fact that these weapons are not treated as particularly different in what little surviving material does discuss them argues against this hypothesis, but only weakly. After all, it is hardly beyond the realm of possibility that one martial arts tradition might treat the warhammer as nothing more than a blunt sword while another might treat it as a unique weapon with its own idiosyncratic method of use.

For myself, though, I tend to think it more likely that no such treatises ever existed. I spoke in Chapter 2 about the inherent limitations of using experimental archaeology to reconstruct historical martial arts. That warning definitely applies here. But—fully acknowledging the limitation of using sparring alone to infer historical weapons usage—my experience has been that fencing against an axe or bludgeon is not very different than fencing against a sword. There is the risk that your opponent will throw his heavy, compact weapon, of course, which can be a nasty surprise. But throwing is one of the attested techniques for these weapons, and one that our limited evidence emphasizes. There is also the risk that a hammer (or axe) can hook a neck or limb in a way that swords can't. But that, too, is covered in even our limited evidence. Other than these things, axes and bludgeons handle very much like short-ranged swords.

To say more than this would be to exceed the bounds of our available evidence. But this is not to say that medieval treatises teach nothing but sword fighting. It is time now to turn to the major class of non-sword weapons that they do cover, the largest and mightiest of medieval hand weapons: polearms.

CHAPTER TEN

Arm's length: thrusting polearms

> *All you need is an arm's length of advantage or more between your weapon and his—and this will keep you safe.*
> —Antonio Manciolino (Manciolino, 2010, p. 145)

So far, the weapons we have discussed in this book have all been some manner of sword or dagger. As the devoted student of medieval weaponry is aware, however, none of these weapons played nearly as important a role in medieval warfare as their larger cousins, the polearms. The time has come at last to discuss these, the largest and most powerful hand weapons of the Middle Ages.

Sidebar: Spear or polearm?

One of the minor but persistent points of confusion I encounter is the question of whether a spear is a polearm—or, to put it another way, just how many weapons the term "polearm" covers. Spears, after all, can be so simple—no more than a pointy stick—while other kinds of polearms can sport heads in quite complex shapes.

> As with all questions of weapon terminology, there is no single definitive answer for all contexts. However, the distinction most often made among historical treatises is not the shape of the iron on the stick but whether or not a weapon has a stick to begin with. In historical fencing circles, the term "polearm" is most often used to encompass any weapon with a long wooden haft, from simple thrusting spears to two-handed axes to elaborate combination weapons with multiple hooks, spikes, blades, and hammers.
>
> When I do see the terms "spear" and "polearm" separated from each other, it tends to be in a gaming context, where the game designer is laboring under the impression that thrusting polearms ("spears") are used in a materially different manner than are chopping polearms. As we will discuss in Chapter 11, there is *some* truth to this, though the overlap between "thrusting" and "chopping" polearms is greater than is often supposed.

In this chapter, we shall discuss what I think of as the thrusting polearms. Fencing treatises have a great many names for these, reflecting variations in the exact length of the shaft, the breadth of the "iron" or head, whether the iron had wings or other lateral projections, and other such nuances. Some of the names for these weapons in historical treatises include *ginetta*, *glefen*, *lancia* or *lanza*, *partigiana* or *partisana*, *picha*, *sper*, and *spiedo*. You can probably guess at the cognates for many of these, but the lexical boundaries of cognate terms rarely line up neatly across languages (for instance, Bolognese use the term *lancia* exclusively to refer to an infantry weapon, while the English term *lance* usually connotes a cavalry weapon). For present purposes, I shall use the generic term "spear" to refer to any polearm whose iron consists of a thrusting point. These range from "short" spears as tall as a man's upraised hand, to long spears in the 9 to 12-foot range (such weapons being especially suitable for use by mounted men, for which reason I shall occasionally refer to them as "lances"), to extremely long spears greater than 12 feet in length (for which weapons I shall occasionally use the term "pike").

Spears sometimes get short shrift in popular media. I have frequently heard, either from new students or from members of the public at demonstrations, confident assertions that the spear was a "peasant's weapon," or that the sword held greater symbolic significance than the spear because the spear was so simple while the sword was rare

> **Sidebar: Fencing with pikes**
>
> I find that students of historical warfare who are not students of historical *fencing* are often surprised to learn that fencing with pikes was a thing. Yet our sources are unequivocal. Marozzo and Manciolino both devote as much attention to fencing with the pike as they do to fencing with other polearms (Manciolino, in fact, devotes *more* space to fencing with the pike than with any other single polearm). Consequently, I make no special distinction in this chapter between fencing with a pike and fencing with a shorter spear, unless explicitly noted.
>
> If this is surprising to you, bear in mind that fencing is not all about—or even mostly about—fancy techniques or "special moves" (for more on the role of fencing in warfare, see Chapter 13). The bulk of fencing training concerns how and when to execute simple actions strongly, quickly, and reliably—the skill to handle even otherwise cumbersome weapons (like pikes), the ability to land firm and accurate thrusts, and a judicious sense of timing. All of this would be relevant to any context in which a man might use a pike, including warfare. Consider the perspective of a pikeman himself: if you might have to stand in the line of battle with a pike, facing enemy pikes, wouldn't *you* want such training?

and precious. When I was young, my roleplaying parties were full of heroes armed with swords, axes, and even hammers, but nobody wanted to play a character armed with a *spear*. Even the lance, a theoretically knightly weapon, was looked upon askance.

All of this appears to be a purely modern development. The anonymous *Ordene de Chevalier*, a highly influential poem describing the ceremony of knighting, describes the spear as a weapon of great symbolic significance, representing a knight's devotion to Truth. Armizare (and it should be remembered here that Fiore was a knight, albeit a minor one) likewise refers to the spear as a "noble" weapon:

> *I am a noble weapon, Lance by name:*
> *In the beginning of battle I am always used.*
> *And whoever watches me with my dashing pennant*
> *Should be frightened with great dread.*
> *And if in the beginning I make my due,*
> *Axe, sword, and dagger will I upset.*

(Fiore Pisani-Dossi, 2019)

While in KDF it is one of the four principal weapons of a young knight:

> *Be a good grappler in wrestling;*
> *Lance, spear, sword, and messer*
> *Handle manfully,*
> *And foil them in your opponent's hands*
>
> (Von Danzig, 2010, p. 96)

Manciolino says that such weapons "can be used with just as much grace and produce the same excellent results" as the sword (Manciolino, 2010, p. 141).

More pragmatically, as Fiore alludes to in the quote above, spears are frighteningly powerful weapons. In modern HEMA, far from being a "peasant's weapon," spears are more likely to be the preserve of advanced students only. A training sword is a relatively safe implement that can only threaten a limited volume of space even in careless hands. A training spear—even one that simulates a short spear—is a much longer lever arm, and can produce so much power that a beginner is likely to injure training partners by accident.

To the extent that we can reconstruct the actual curricula of medieval martial arts schools, this seems to have been the case in the past as well. While Liechtenauer's zettel refers to lance and spear in its introductory verses, it does not actually discuss the use of such weapons until after his longsword section, which is the largest of the zettel's three parts. In Fiore's treatise as well, use of the spear is taught last of all the foot weapons: after grappling, the marshal's baton, the dagger, the longsword, and even the poleaxe. Marozzo, meanwhile, explicitly advises the fencing teacher to begin new students with the sword, and both he and Manciolino—the only Bolognese masters to discuss the use of the long spear or pike—discuss it last of all in their respective texts. While it is impossible to prove that fencing masters taught their students in the order that they arranged their texts, these facts are evocative. Moreover, there is a noticeable lack of any suggestion in our extant treatises that medieval martial artists held the spear in anything other than high regard.

One reason for this is that spears hit very hard and very fast. I find that this often surprises non-fencers, though I am not entirely sure why. Perhaps, like me, they grew up with games telling them that swords and axes do more damage than spears. Perhaps it is simply a matter of visual intuition: something about the all-steel sword just seems more menacing than the mostly-wooden spear. Whatever the reason, it is a misconception that is easily cured with a few sparring sessions.

One of the factors behind a spear's apparent speed is the way in which it can thrust. A spear can thrust like a sword, extending first the arms, engaging the core muscles, and finally moving the feet. It can also thrust in a pool-cue fashion, which tends to conceal the movement of the rear hand. The spear itself does not necessarily move any faster than does a sword, but the greater variety of methods whereby it can thrust mean that there is more for an opposing fencer to watch out for. A spear fencer can use this to steal a mental march on the opponent, which creates the illusion of great speed. Learning to exploit an opponent's momentary lapses or breaks in attention in order to create the illusion of speed is an essential skill for any good fencer; spears simply add another tool to do with thrusts that is not available to most weapons.

One characteristic of spears that is not an illusion is the power behind their attacks. To put it frankly, spears hit like a freight train. Even a small medieval spear can easily weigh 4 pounds, which is more than all but the weightiest swords. The form factor of a spear also makes it very easy for a fencer to put the entire weight of the body behind the spear thrust, braced against the body with a solidity that it takes great skill to replicate in a sword thrust. Spears can deliver quite significant percussive blows, as well—while their irons may not be optimized for anything but thrusting, a solid weight of steel at the end of a pole 6 feet or more in length packs an enormous wallop (far greater, in fact, than the percussive blow of a sword pommel, even one that is swung with both hands on the blade). Spears are enormously powerful weapons.

What, then, can be done with such mighty weapons? The first answer is another aspect of historical spear fencing that often surprises people: throw them.

Of course, if you are reading this book, it is likely not a surprise to you that certain types of spears can be thrown. Spear throwing is even an Olympic event, in the form of the javelin toss. What may surprise you, however, is the sheer variety of spears that can be thrown. It certainly surprised me to learn that medieval fencers were expected to be able to accurately throw their lances! Yet the very beginning of Liechtenauer's armored fencing on foot begins with a combatant making the choice to dismount against a superior rider, continuing the fight with his lance on foot. Here is the verse from the zettel, with gloss by Pseudo-Peter von Danzig:

> *Whoever dismounts, begins to fence on foot*
> Gloss—Note: this is a lesson: one can fight in harness on foot and on horseback ... And if it should happen that you must strike or fight with an opponent on horseback, and you realize that he is too clever or too strong at it, then dismount to go on foot.
> (Von Danzig, 2010, p. 144)

Because mounted fencing is first and foremost a contest of equestrianship, a mounted fencer may well be overmatched by a superior rider despite superior skill in weapons play. As this commentary makes clear, KDF teaches that there is no shame in the inferior rider dismounting if one's mounted opponent is "too prudent or too powerful" on horseback. Dismounting allows the inferior rider to rely on skill at arms.

Of course, a dismounted combatant will still be armed for cavalry combat, and obliged to fight with a lance on foot. As the following commentary by the actual Peter von Danzig makes clear, though, even a lance may be thrown:

> *Spear and point, thrust the initial thrust without fear*
> Gloss—Note: this means when you have placed or positioned yourself to throw your spear, then cast the throw with care. If you hit, then follow after the throw.... Now note that if you don't truly hit the opening with the thrust or throw, then do not fall to the spear, so that you do not lose your balance and he throws you down, but grasp your sword.
> (Von Danzig, 2010, p. 192)

Von Danzig's main concern here is to advise the fencer not to rush in too hastily on the heels of his thrown spear should his cast miss the mark. Note, though, the implicit assumption: that a dismounted combatant can hurl his lance with such force and accuracy that it *does* "truly hit the opening" of a plate-armored opponent.

I should hasten to add that when medieval fencing treatises discuss mounted lances, they do not generally seem to refer to the great counter-weighted cone-shaped lances one sees at jousting reenactments.

The lances for dueling or war that our illustrated manuals depict are generally much simpler affairs, stout sticks with pointed irons, and it is reasonable to assume that Liechtenauer and his glossators were speaking of a similar type of lance. It is also worth noting that von Danzig assumes the throwing fencer can "follow after" his throw to exploit the opening it creates, so the range for this technique is quite short—very nearly within thrusting range, if modern experiment is a guide. Nonetheless, hurling a 12-foot spear into the gaps between plates of an armored opponent, even from just a few yards away, is an impressive martial feat.

Spear throwing is absolutely ubiquitous in medieval treatises. In addition to the KDF example we have already seen, Fiore discusses four separate scenarios involving thrown spears, one of which involves an opponent who has brought two spears to throw and two of which involve being attacked by multiple opponents at once, including one or more who throw weapons. Both Marozzo and Manciolino include dedicated sections on the topic, including drills on how to throw multiple spears and how to defend against an opponent who has multiple spears to throw. Marozzo also includes instructions on how to beat a thrown spear out of the air with a two-handed sword, "as I have put to the test on numerous occasions" (Marozzo, 2018, p. 265). Monte, meanwhile, includes no less than eight separate sections in the *Collectanea* on throwing spears of various types, on foot and mounted, including suggested remedies for those who injure their arm practicing their spear throwing too much.

It is difficult to dismiss all of this as mere sport or martial exercise, particularly as both armizare and Bolognese include multiple-opponent scenarios involving thrown spears. At the same time, I would forgive you if you were a little bit skeptical that medieval combat consisted largely of fencers volleying spears back and forth like Homeric heroes. Nor is that the suggestion of our sources (though our 16th-century authors were likely aware of, and pleased by, the similarity). To return to our first example, von Danzig immediately follows his instructions on throwing the lance with the gloss of another verse, writing "This is if you want to remain and work with the long weapon or with the spear using the thrust, and you don't want to throw the spear" (Von Danzig, 2010, p. 192). In other words, the dismounted fencer has the choice to throw or not to throw, and von Danzig expresses no opinion as to that choice. Manciolino likewise introduces his first spear-throwing technique with the words "if you both *decided* to cast the partisans at one another" (Manciolino, 2010, p. 141, emphasis added).

No doubt in many instances—probably in most instances—it would be unwise to throw one's spear. One of my students once misunderstood me on this point, and imagined lines of medieval soldiers marching up to each other and promptly flinging their spears away *en masse* before closing with swords, and I certainly do not mean to suggest that (which is not to say that it couldn't work—a mass of thrown spears followed immediately by close work with swords was the signature tactic of Roman warfare for many centuries)! Rather, I wish to emphasize that spear throwing was always an option: a threat to beware, and a technique to keep in a fencer's back pocket for moments when a thrown spear could make all the difference.

In a HEMA context, thrown spears greatly increase the danger inherent in spear sparring. A skilled fencer can control the force of a spear thrust in several ways without much compromising its speed, but a spear *throw* must have a certain amount of force behind it lest it be unrealistically easy to defend. Moreover, a missed throw can pose a serious safety hazard to bystanders—other students, competition staff, or audience members.

Nevertheless, it has been my experience that permitting throws (with appropriate extra safety precautions, of course) does change the dynamic of a spear fight in important ways. In many such matches, no spear is thrown, but the fact that one *might* be exerts a noticeable influence on the fencers' behavior. When spears can be thrown, a fencer must be on guard much further away from his opponent than otherwise. In a skirmish (one of the contexts that Marozzo specifically discusses), throwing permits a spear fencer to support his allies from further away than he otherwise could, adding to the potential threats the other side must keep track of.

Even in a one-on-one context, the possibility that a spear might be thrown means a fencer must add to his mental load by keeping track of signs that his opponent is about to throw. A fencer can add to this by holding his spear in an overhead grip, adding an element of uncertainty as to whether he will attack with throw or thrust.

Thrown spears also add uncertainty to a fencer's distance. If spears cannot be thrown, a spear fencer can only initiate a rush to close range from close enough to shove the opponent's weapon out of position. If spears can be thrown, then a fencer who is hovering just outside of thrusting range may in fact be preparing to hurl the spear and close with the sword—which, again, gives a spear fencer more possibilities for misdirection.

Exploiting this capacity for misdirection is a major component of historical spear work. Of all the medieval weapons for which we have significant surviving technical material, spears have the fewest techniques in the formal sense. "There aren't very many blows used with pole arms," says Marozzo, "since one generally throws almost nothing but thrusts" (the wide-bladed partisan can be used to cut, he adds, "but only rarely") (Marozzo, 2018, p. 277). Despite this, they are among the most formidable weapons a fencer can face, for they have many ways to apply their limited universe of techniques. Spears are not complex weapons, but they *are* extraordinarily subtle.

It will be worth it to quote another section of Pseudo-Peter von Danzig's commentary on the zettel to illustrate this point. In the scenario that follows, both fencers have dismounted and fight with spears in armor:

> *Spring, wind, truly plant upon him.*
> *If he defends, pull, that will defeat him.*
>
> Gloss—*Note*: this is another: when you have both dismounted and each has his lance … then hold it by your right side in the Low Guard and go toward him in this manner. And thrust skillfully to his face with outstretched arms. If he thrusts to you at the same time, then go up with your lance against his into the High Hanging, and with this, spring to him and plant upon him above. If he goes up with his arms and defends against the High Thrust, then pull and set your point to him under his left armpit at his opening or wherever you can and press him forward.
>
> (Von Danzig, 2010, p. 146)

This is a three-part play. The fencer begins by thrusting from the lower right at the opponent's face or eye slits (for more about targeting "openings" in plate armor, see Chapter 12). One method for the opponent to defend against this thrust is to thrust back so that the two spears collide and the first thrust is deviated from its path (the eye slit of a helmet, after all, is not a large target). The attacking fencer can counter this simply by raising the hands to avoid the collision or even to snake the point back on target. If the opponent's spear raises in a more exaggerated parry to push the attack out of its path yet again, the attacking fencer can jerk the spear back, away from the opponent's defending spear, and slam a new thrust home into the opponent's armpit.

This same play can be performed with longswords held at the short sword (with one hand on the blade), which presents an interesting contrast. When performed with swords, it takes place at very close range—so close that the fencers' armor is likely to scrape against each other. This same short range necessitates fairly large movements of the arms in order to achieve the required angles. Spears, however, by virtue of their length, can produce very large movements of the point through very small movements of the back hand. Thus, performing this play with spears is an entirely different physical experience. The arms and body must still be moved, of course, but compared to performing the play with longswords, the fencers seem almost still as their spear points whip through the air.

Many of the fundamental spear techniques are on display here. As Marozzo implies, they are not many: the spears thrust and change their attack angles by winding around each other and pulling back to escape parries. An example from Manciolino will help to illustrate some more. In the example that follows, the fencer begins with the left foot forward, partisan pointed forward and to the left:

> Let's suppose that the opponent is the first to strike, with a thrust aimed at your leg. Parry the blow by hitting into his incoming weapon towards the outside (that is, towards the opponent's left), with your right hand high and the iron of your partisan pointing down. Then, riposte with a thrust of your own to his flank or leg (as you prefer), retreating to safety with a rearward jump ... If the opponent attacks you with a thrust to the face or with a cut, push your right hand down so that the iron of your partisan ends up in front of the opponent's face. By doing this, you will defend from his attack. Then, riposte with a thrust to his flank.
>
> (Manciolino, 2010, p. 142)

Here, Manciolino uses the ability of a polearm to perform large actions with small inputs to invite the opponent to attack the leading leg. The defending fencer then uses the long lever arm of the spear shaft to whip the weapon into a forceful parry, knocking the attacking spear to the defender's right. If the parry connects, the defender can return the favor with a thrust to the opponent's leg. Of course, the attacker might also attack high, or even feint low and then attack high—in which case, the

defender can snap her own spear up by pushing down with the rear hand and snapping the butt of the spear to the hip, again pushing the incoming thrust to the defender's right, before counterattacking. Here, the long lever arm of the spears is used to produce a staccato series of attacks and parries high and low, with a speed that a weapon with a shorter lever arm would be hard-pressed to match.

Note also the importance of both offering and attacking the leg. The leg may seem like a small target for a spear thrust, but precise thrusts to the legs and even feet are an important part of historical spear play. There are several reasons for this. The first is that, as we just saw in the play from Manciolino, a spear fencer facing another spear must be ready to change targets high and low to land a thrust. Second, in a mixed-weapon context, the reach advantage of a spear allows a fencer not just to strike the opponent from further away, but also to strike different *parts* of the opponent while remaining out of reach. If a spear fencer were restricted to thrusts above the belt, the weapon would be much more predictable and thus less threatening to the shorter weapon.

Finally, in an armored context, the sabaton (foot armor made of articulated lames) was one of the few plate targets that stood a reasonable chance of being penetrated under fighting conditions. While it is usually axiomatic that plate armor is proof against penetration in live combat, armor plate can be penetrated when the plate is braced and steadied against a surface so that the incoming blow almost crushes it against the surface. In fighting conditions, this usually does not occur—but a thrust to the foot, by its nature, crushes the foot armor against the ground, making penetration possible (it helps, of course, that the lobstered plates of a sabaton cannot be of any great thickness and retain articulation).

For all these reasons, a spear fencer who is not accurate enough to hit the legs and feet surrenders a significant amount of the weapon's versatility. Manciolino even makes oblique reference to a scoring system for sparring in which a hit to the foot scores two points (while hits to any other parts of the body except the head score one), "in deference to the awkwardness of delivering such a low attack" (Manciolino, 2010, p. 78). In short, low thrusts might be difficult, but they are an important skill for a spear fencer to develop nonetheless.

Before we move on to the next set of spear techniques, we should spend a little more time on the subject of percussive blows with the spear. We have already seen one example of such movements in the Manciolino play. There, a percussive blow of the spear was used to forcefully parry an incoming thrust to the leg. Hard blows with the shaft of a spear are not common in all forms of European fencing—KDF, for instance, largely eschews them—but they are an important topic in medieval spear play as a whole.

Even when they are used, percussive spear blows tend not to be attacks. A hit from a spear shaft can deliver quite a bit of force, but it tends to leave the fencer unnecessarily vulnerable (a topic we will discuss in more detail in Chapter 11, when we discuss the "heavy" polearms), and it is generally not as injurious as impaling an opponent on a pointy stick (a topic that will become especially relevant in Chapter 12, when we discuss armored fencing).

Instead, these techniques are most often used defensively, to smash an incoming weapon aside and create an opening for a counter-thrust. Armizare's spear work relies heavily upon this basic technique, even going so far as to incorporate multiple stances in which the spear is held over the shoulder almost like a baseball bat.

In the armizare mindset, this defensive use of the spear is especially desirable because it works even when the opponent has a reach advantage (not all spears, after all, are of equal length). But I want to dwell on it because it helps to illustrate other features of spear fencing as well. The following armizare play will introduce the next two important concepts:

> I'm positioned in the left side Window Guard [fenestra sinistra]. ...
> I'll step offline to the left with my left foot, with my point held high and my arms low. Then I'll thrust into your face and you'll have no defense.
>
> (Fiore, 2017, p. 40r)

In this play, the defending fencer is positioned nearly with the back to the opponent, spear over the shoulder held in crossed hands, offering the right side as a tempting target. When the opponent takes the bait, the fencer pivots around the right foot, stepping around with the left, and uncrosses the hands, which powers a strong beat with the spear to push the opponent's thrust aside. The play finishes with a thrust to the face.

The pivot turn, beginning from such a coiled-up starting position, is a good example of something we have not yet touched on: good spear fencing is not always a static affair of small arm motions and snaking spear points. True as may be the points we have made about the use of the spear as a long lever arm, it is also true that spear fencing involves active, even dramatic footwork. Fiore's uncoiling turn is one example. Recall also the "rearward jump" from the Manciolino play we have already discussed. In this play, from Manciolino's use of the pike, Manciolino calls for a run of "four or five steps" to out-angle the opponent:

> Feint a thrust, which he will try to avert by passing back with his left foot. At that point, run four or five steps obliquely towards his left side: thus, finding him open, you will be able to push a thrust into his flank.
>
> (Manciolino, 2010, p. 145)

Marozzo includes a particularly dramatic pivot turn in his own pike fencing, in which the whole body is whipped like a snake to power a rising parry with the heavy long weapon and it is then whirled overhead. The play begins with the fencer offering his left flank to the opponent, pike drooped toward the ground on his right side:

> So let's propose that he does attack your aforesaid left flank. You will then raise your arms into the air, both of them above and behind your head, beating his blow behind your back with your spear, bending at your back and leaning your head back and your belly forward, and thereby you'll parry his thrust.
>
> (Marozzo, 2018, p. 280–281)

The overhead rotation of the pike is not mere flash; it facilitates the fencer changing lead hands so that when the pivot turn is finished, the pike is now on the left side, between the fencer and the opponent's weapon. This whirling rotation is not, we may presume, one that a pike fencer would be expected to apply in formation, but the larger lesson of powering defensive beats with the spear shaft by fully engaging the whole body is applicable to all spear fencing. I can assure you from my own experience that, properly powered and timed, such defenses can literally knock a spear from an opponent's hand.

> **Sidebar: Dancing with spears**
>
> In speaking of "dramatic" spear footwork, I do not wish to be understood as suggesting that medieval spear fencers performed gymnastic maneuvers such as one might find in a martial arts film. This is doubly true for armored spear fencers; indeed, as we will see in Chapter 12, armored fencers were specifically advised *not* to move too much or unnecessarily, lest they exhaust themselves. However, neither should we fall prey to the opposite extreme, often presented in popular media, that spear fencing consisted of nothing but two men (or lines of men) squaring up to their opponents, projecting their spears forward, and stolidly stabbing away. Real spear fencing is considerably more active and mobile, even if only in the turns of the body, than *that*.

What is a spear fencer to do against such powerful percussive beats? There are a number of options. We have already seen one, in the very first spear play we looked at this chapter, from KDF. Recall that in that play the attacking fencer counters her opponent's second, more exaggerated parry by pulling her spear through her lead hand. In this manner, the opponent's beat finds nothing but air, and the attacker is all the more primed for a powerful thrust to the other side of her opponent's spear as it goes by. Another canonical KDF counter would be to use a lever action of the rear hand so that the attacking spear dips or circles below the percussive parry, again causing it to miss. KDF spear work is particularly in love with such actions, perhaps precisely because it tends to eschew the more powerful beats of which armizare and Bolognese make liberal use. From the KDF standpoint, beats with the spear are often needlessly exaggerated and even dangerous. From the Italian perspective, this is only true if they are improperly timed.

If a percussive parry *does* connect, Fiore offers a remedy. The counter to his own uncoiling technique, he says, "is easily done: when your thrust is beaten offline, you turn the butt of your spear and strike with that" (Fiore, 2017, p. 40r). It is wise, he says earlier in his book, to cap the butt of your spear with well-tempered steel for just such an occasion.

This brings us to another major point in the medieval use of the spear: if a spear *is* knocked aside, the canny fencer can accept that momentum and simply rotate the spear to continue fencing with the other end the weapon.

There is something viscerally satisfying about a well-timed butt strike, but it is easy to over-use strikes with the butt of a spear in modern sparring. Striking with the butt of the spear is a perfectly sensible historical technique in the case that Fiore describes, where the choice is between losing control of the spear due to a well-timed percussive parry and rotating the spear to parry an incoming counter-thrust or strike a charging opponent. However, it is easy for modern fencers to use it as a generalized infighting technique, and this is not in accord with historical practice. Context, and the way we modern fencers conduct our sparring games, is important here, for the preferred methods of fighting up close with a spear in historical treatises are (i) with wrestling or (ii) with a sword. Modern fencers really have no excuse not to be wrestling when up close with their spears, but the reason not all HEMA fencers draw their swords is simple: not all of us who spar with spears *wear* swords.

Historically, however, the ability to transition from spear to sword was important. Von Danzig shows a preference for grappling rather than using the sword when spear fighting comes to close range, but he also recognizes the danger of the *opponent* transitioning to a side-arm, and warns the student not to let that happen. In the play that follows, von Danzig assumes that a fencer has managed to grab the opponent's spear:

> [W]atch for his drawing of the dagger or messer or sword before he draws, and don't let him do it, but grasp him in front of his hand or his body, and push him down with skill as was described in the wrestling, so that you come near him; thus he cannot harm you with his sword, either by planting upon you, or by striking with the pommel.
>
> (Von Danzig, 2010, p. 194)

Manciolino describes another example of a sidearm transition, this time with pikes. In this amusing technique, the attacking fencer deliberately throws his pike athwart his opponent's to weigh it down, then charges to close-range work with sword or dagger:

> [M]ake him parry ... As he hits your spear, cast it at an oblique angle and let go of it; it will fall on his weapon towards his right side. Use that tempo to run against the butt of his spear, then unsheathe your sword or dagger (which you will be wearing at your side) and, reaching him thus unexpectedly, you will be able to strike him at will.
>
> (Manciolino, 2010, p. 144)

The Gladiatoria Group gives some sense of how important it was for a spear fencer to be able to swiftly transition from spear to sword. Of the 12 spear plays in the Gladiatoria, fully one-quarter (the fifth, sixth, and seventh) techniques describe scenarios in which a fencer has thrown the spear but missed, and must now deal with a spear-armed opponent while momentarily unarmed. In the fifth play, the fencer beats away the incoming spear thrust with a sword cut from the draw (not an easy feat to perform under pressure, by feel alone, in gauntlets!). If the defending fencer does not have time even for that, the sixth play recommends a draw of the dagger instead, parrying the incoming spear thrust with the dagger. The seventh play teaches the fencer to beware, for even if the opponent's spear thrust is defeated, the opponent may immediately transition to the sword as well (particularly worrisome if the defender had to draw the dagger rather than the sword). Together, these three plays give a sense of the rapid transition between spear and sword that a fully armed spear fencer was expected to be able to perform.

Again, none of this is to say that the spear was helpless against a charging opponent. Spears are quick to retract, after all, and spear fencers themselves can move backward. Many a modern fencer has discovered that "just parry/grab the point and rush in" is easier said than done. Neither is the spear completely helpless even at close range; it can still be used to batter or cross-check an opponent away, hockey stick-fashion (this is, in fact, a documented close-range technique with the poleaxe). However, there is no particular *reason* for a spear fencer to continue attempting to fence with the spear against an opponent who has managed to close rather than draw either sword or dagger.

The last topic in spear fencing we should discuss is the use of spear and shield together. Bolognese is the only extant system that treats this particular weapon combination, although the fact that Monte discusses it as well suggests that it may have commanded more attention from men of the martial class than we might otherwise infer.

In both Monte and Bolognese, the shield used in spear and shield is the rotella, the slightly domed, round, arm-strapped shield that occupied most of our attention in Chapter 5. The spear paired with the shield is universally the partisan, which is the smallest of the Bolognese spears ("smallest" being a relative term, of course; Monte describes the partisan as a little bit taller than a man's upraised hand).

The first thing to say about spear and shield fencing is that it is physically taxing: this weapon combination puts more weight on the fencer's arms than any other combination in this book. This is exacerbated by the fact that the addition of a shield to the normal equation of spear fencing puts extra demands on a fencer's footwork. Sprightly footwork and careful attention to the angle of engagement is especially necessary in order to overcome the defensive barrier of a well-handled rotella. Marozzo illustrates this principle well in the following sequence:

> Now, having your left leg forward, I want you to extend your shield toward your enemy, and without taking a step, attack his forward leg with your partisan.... If he responds with a thrust to your face or leg, step toward his right side with your right leg, and as you step, use the haft of your partisan to knock his blow toward your left side, and throw a punta riversa to his chest, between his shield and his partisan.
>
> (Marozzo, 2018, p. 271)

In this play, the fencer provokes a thrust by attacking the opponent's leg with an overhand thrust, holding the partisan so that the thumb of the spear hand is toward the butt of the weapon. When the opponent counterattacks, the fencer crosses the right leg in front of the left, twisting the whole body to the left, while simultaneously dipping the head of the partisan toward the ground to parry the incoming thrust with the haft, moving the haft across his body like a wooden wall. This coils the fencer's whole body to the left while simultaneously moving toward the opponent's spear side and pushing the opponent's spear out of position. Having thus momentarily achieved an angle against which the opponent's shield cannot cover, the fencer uncoils with a backhanded thrust to the opponent's chest.

Many of the elements of spear fencing we have already seen are present here, such as the importance of the leg as a target, the need and ability to switch between high and low targets (the attacker threatens the leg and ultimately strikes the chest; the defender is assumed to be able to counterattack either face or leg with equal ease), and the powerful coiling footwork. One item that is less present is reach: a partisan held in one hand must be held at or near the midpoint of the haft or the weapon becomes unmanageable. And there is another feature that is absent entirely, which is the use of the spear as a lever arm.

You can probably already guess at the reason why: levering a spear requires one hand to act as a fulcrum and the other to manipulate the lever, and a spear and shield fencer does not usually have two hands on the spear. The spear is usually held overhand, while the non-dominant hand grips the shield.

This has significant consequences for the spear's ability to parry. A spear held in two hands can often parry an attack simply by thrusting at the opponent along the correct vector, as we saw at the very beginning of this chapter. This keeps the spear point close to the opponent, which facilitates the defender's counterattack. A spear held in one hand, though, is significantly less stable. A spear held in one hand can only parry effectively by turning its iron toward the ground and making a coarse sweeping motion left or right.

While the loss of lever action is a negative, it does come with compensatory advantages. For one thing, the shield can (and should) be used to protect the spear hand from attack, while in the standard two-handed grip the fencer must always be wary of quick attacks to the forward hand. For another, the standard overhand grip on the spear allows a spear and shield fencer to throw the spear with very little warning, as it is essentially always primed to throw. Lastly, the shield is free to be an active weapon. The shield is frequently used to beat aside incoming thrusts, leaving the spear free to attack. The shield, of course, can attack as well. As we discussed in Chapter 5, this most often takes the form of suppressing the opponent's weapon arm. In this play from Marozzo, the student uses the partisan to knock the opponent's weapon to the right, creating an opening for the shield to attack in just such a manner:

> [P]ut your right leg forward and offer your exposed right flank to your enemy, with the point of your partisan turned toward the ground. Do so in order to give him cause to attack your right flank, and, knowing that he will do so, send your left leg well forward, toward his right side. As you step, knock his blow outwards with your partisan, toward your right. In the same tempo in which you step and parry his blow, stick your shield into his right arm in order to throw a gripped punta dritta to his chest in such a way that he'll be unable to move his own partisan, since you'll have bound him up with your shield.
>
> (Marozzo, 2018, p. 271–272)

As in Chapter 5, notice that the blow of the shield to the arm is merely a preparatory attack for a blow with the bladed weapon—in this case, a thrust from the right with the spear held so that the knuckles face the ground.

You may have noticed a commonality in the plays we have discussed thus far: both end up attacking the opponent from his right side (i.e., his spear side). This is no coincidence. The main targets in spear and shield fencing are, unsurprisingly, the parts of the opponent that are hardest to defend with the shield: the legs and feet and the spear-side arm and flank. The face runs a distant third-priority target, targeted mostly as a way to force the opponent's shield to commit to a defensive action.

This is not a large number of viable targets. As a result, spear and shield fencing is often quite a protracted affair compared to the pace of fencing with spear alone. Monte gives a good sense of this careful waiting game, warning that when fencing with spear and shield, a fencer must enter and exit distance "moderately and often" (Monte, 2018).

Spear and shield fencing does not have to use the two weapons separately. One of the advantages of an arm-strapped shield, as opposed to a center-gripped one, is that the straps can be loosened so that the shield hand peeks beyond the shield rim. This allows the fencer to grip the spear in two hands while retaining some protection from the shield: the spear is held on the right side of the body, with the shield to its left, covering the chest.

This stance should be seen as a midway point between holding the spear in two hands without a shield and holding the spear and shield separately. It does provide additional protection to the fencer's chest and left arm compared to fencing two-handed without a shield, but it lessens the mobility of the shield on both offense and defense and also prevents the spear from ever being held on the left side of the body (without a shield, as we have seen, a fencer can swap leading hands to swap the spear from one side to the other, which opens up both offensive and defensive possibilities). Within those constraints, however, a spear and shield held in this way operate essentially identically to a spear held in two hands. Indeed, the techniques in Manciolino's section on this manner of holding the spear in this manner are exactly identical to the techniques in his section on use of the partisan alone. Discussing this manner of holding the spear and shield, Marozzo candidly admits, "[Y]ou would not be at any disadvantage if you were to cast away your shield and hold your partisan gripped with both hands" (Marozzo, 2018, p. 272–273).

There are two final topics in sword and shield fencing we should discuss. The first has to do with fencing with spear and shield against armor. One of the disadvantages to fencing with spear and shield, compared to spear alone, is that it becomes much more difficult to solidly plant a spear point in mail when it is driven with one hand—and, once planted, it is much more difficult to drive an armored opponent before it. A two-handed thrust is simply much more structurally solid than a one-handed thrust, particularly as a single hand holding a spear will be very far from its point. Against an armored foe, a spear and shield fencer would be very grateful for that the rotella gives him the ability to grip his spear with two hands.

This brings us to our last point, a delightful curiosity from Monte that may be more serious than it first appears. He writes that, when a fencer is armed with a corselet, spear, and rotella, an exhausted fencer can fling the rotella at the opponent's face, following with the spear to finish the fight:

> If one has a corslet together with a *rotella*, our *rotella* can sometimes be thrown at the other's face, especially when we are fatigued in any way. And at that moment one must reach with the partisan, held in both hands, since, while he is engaged with this impediment, we can wound him.
>
> (Monte, 2018)

Yes, you read that right: a spear and rotella fencer can fling his shield at an opponent's face.

While this technique has pop culture resonances in the present day that Monte could never have anticipated, it has serious martial applications, as well. It is, as he says, a method of discarding the rotella productively when a fencer is exhausted (not an improbable scenario for a heavy weapon combination that lends itself to protracted, cautious fencing). It is also a method of discarding the rotella to attack with both hands on the spear against an armored opponent, one that does less to disarm the fencer than throwing the spear and attacking with the sword.

Speaking of armored spear fencing, it is also not a coincidence that Monte recommends throwing the shield only "if one has a corslet"; i.e., armor for the torso. In this chapter we have looked in detail at how the seemingly uncomplicated spear is actually a wickedly deceptive, subtle, fast, and powerful weapon. Although Marozzo explicitly notes that polearm fencing is sometimes performed without armor, unarmored spear fencing is, in my candid opinion, murderous lunacy (I also think it is a great deal of fun, but it would not be so if the weapons were real). In any ordinary combative circumstance, a fencer facing a spear would be well advised to obtain protection of some sort—if not a shield, then at least a corslet.

Which would win: spear and shield and sword and shield?

Both the spear alone and the spear and shield are a very difficult weapon for any smaller weapon to face. Even greatswords, as we will see in Chapter 11, have difficulty with polearms. Armor or no armor (provided both combatants are equally armored or unarmored), a fencer with a sword against a fencer with a spear is at a significant disadvantage. Spear and shield vs. sword and shield, however, is a much more even contest.

Marozzo treats the use of sword and shield against polearms generally (without specifically discussing spear and shield). Notably, he does not add the warning that the matchup is unequal one way or the other, which he does add in his discussion of the two-handed sword vs. polearms.

In this scenario, the swordsman's shield significantly mitigates many of the advantages a spear normally has against a sword. Because the shield can protect so much of the body, a spear's ability to deceive and switch targets is significantly reduced. There simply are not as many targets to choose from, and thus not as many targets for the swordsman to worry about protecting. Additionally, because a spear held in one hand has less reach than a spear held in two, the swordsman is not (usually) at as significant a reach disadvantage against a spear and shield as he would be against a spear alone. This further mitigates the spear's ability to switch targets with impunity.

Lastly, a one-handed sword has far more leverage than a spear in one hand. A one-handed sword can push aside a spear in one hand with relative ease. Indeed, this is the core of Marozzo's advice when using a sword and shield against any polearm, even a spear held in two hands: use the sword or shield's superior leverage to push the polearm aside, and close quickly to sword range.

This is not to say (and Marozzo does not claim) that this is a foolproof method, but it is a significant challenge to a spear and shield fencer. Against these difficulties, the spear fencer has the defensive utility of a shield, the ability to throw the spear, and the fact that the spear does enjoy at least modestly superior reach. And of course, if the spear fencer feels that either the modest reach advantage or low leverage of a spear in one hand is inadequate, there is always the ability to hold the spear in two hands, immobilizing the shield in exchange for greater range and leverage.

And yet, regardless of how the spear is held, the spear fencer must contend with the fact that the opponent's shield makes it fairly easy to close aggressively against the spear. In order to maintain the advantage of the spear, the spear fencer must be ready to retreat with alacrity (and must also be well trained in the transition to sword or dagger!). At the same time, the only hits in spear and shield fencing that tend to happen at long range are those to the leg or to the spear arm as it extends—attacks to the body from the weapon side generally require the fencers to be fairly close in order to achieve the required angles before the opponent can react. As a sword fencer has no reason to expose the arm in a futile attempt to intimidate a spear fencer with thrusts (Marozzo notes that the swordsman should keep his shield "close together with your sword"), this means that the spear fencer is likely to have to immobilize the opponent with a thrust to the leg or else get fairly close anyway (Marozzo, 2018, p. 201).

This is the factor that tips the balance in favor of the sword and shield in my opinion. The issue is not free from doubt, and the weapons sets are quite equal, but it is the nature of shield fencing to contract the range of the fight. More often than not, I believe that a contest of spear and shield vs. sword and shield will either end in favor of the sword and shield or force the spear fencer to transition to the sword.

I should add a postscript here regarding the skirmish context. As we have alluded to several times in this chapter, a fully utilized spear performs well in support of friendly fencers. Thus, while I think that a single sword and spear fencer will be able to overcome the spear (if not necessarily the sword) of a single fencer armed with spear and shield, it is likewise my opinion that the spear and shield is a superior weapons combination for skirmish work. The spear can be used from a greater distance to support a friendly fencer or even thrown, and the spear and shield fencers always have their own swords to fall back on.

CHAPTER ELEVEN

Grace and results: cutting polearms

These weapons can be used with just as much grace and produce the same excellent results as the ones we have already seen.
—Antonio Manciolino (Manciolino, 2010, p. 141)

Having discussed stabby polearms in Chapter 10, it is time that we turned our attention to what we might think of as the "heavy" polearms. This chapter is principally concerned with three weapons: the halberd, which has an axe-like blade; the poleaxe, which has some combination of a pick-like beak, axe blade, and hammer face; and the bill, which has a broad, almost sword-like blade with a sharpened curve almost like a pruning hook. These are weapons that have significant capability to deliver blows with a percussive attack.

We must begin with an obvious caveat: *most* polearms have a significant capability to deliver blows with the edge. Even a fairly short 6-foot spear with a narrow-bladed stabbing iron can deliver quite the wallop. The force that such a long lever arm can generate is not to be taken lightly. Plenty of HEMA fencers who train with polearms know of at least one person who has been knocked unconscious even by a "simple" quarterstaff, or by a polearm hit that their partner genuinely thought was a gentle tap.

Despite this truism, "heavy" polearms can deliver blows of truly fearsome power even by polearm standards. A friend of mine once related a story of poleaxe sparring that really drove this point home for me. Early in his HEMA career, my friend and his sparring partner were working with "safe" poleaxes equipped with rubber heads and stout hickory shafts. They were wearing plate armor, and so felt comfortable hitting with quite powerful blows. Mindful that not even plate armor was historically considered proof against poleaxes, they were also taking care to block these powerful blows.

And their poleaxe shafts snapped in half.

Historical poleaxes were frequently reinforced with langets—strips of iron riveted to the shaft—but my friend's example nevertheless drove home to him that when historical treatises discuss the need for finesse with polearms, they mean it.

> **Sidebar: Axes vs. hammers**
>
> One of the questions that bedeviled my teenage self was how axes and warhammers differed from each other in practical use. Imagine my surprise to discover that, from a fencing perspective, there is no difference at all—in fact, they even share a name!
>
> This chapter will make frequent reference to the "poleaxe," a term that stands in for the French *hache*, German *axe*, and Italian *aza, azza*, and *ascia*. All of these terms mean "axe" in the dictionary sense, but as a technical fencing term, they all clearly encompass hafted two-handed weapons with hammer faces—and, in many cases, no blades at all. For instance, Fiore and Paulus Kal both illustrate "axes" that have no blades, only hammer faces and back spikes.
>
> My teenaged self would have objected—vociferously—that such weapons were properly termed two-handed warhammers, or becs de corbin. Our treatises, obviously, would disagree. For this reason, everything in this chapter should be taken to include two-handed hammers as well as two-handed axes.

This finesse is the first thing that must be emphasized for heavy polearms. The first play of the earliest extant poleaxe treatise, *Le Jeu de la Hache*, imagines two poleaxe fencers with the heads of their axes held high. The attacking fencer delivers a downward diagonal blow with the head of the axe. The defending fencer parries not by swinging back, but by sweeping the tail of the axe up and to the right. This merely deflects the incoming blow, and positions the tail of the defender's axe to jab like a pool cue at the attacker's head.

Le Jeu is far from the only source that recommends this use of the butt. Talhoffer includes a very similar play in his poleaxe work. The Anonimo Bolognese likewise pays tribute to this technique in its own first play of the poleaxe, this time advising the fencer how to counter it by deviating the original blow so that the sweeping butt does not connect:

> [B]e aware that your enemy may pass forward with his left foot, beat the heel of his poleaxe into your haft, and follow that by striking your face with the heel of his poleaxe. So in that tempo when he goes to beat your haft, raise your poleaxe high and make his blow go in vain. Then immediately give him an axe-blow to the head.
> (Anonimo, 2020, p. 235)

Neither is this use of the butt end of the weapon to parry unique to the poleaxe. Over a century after *Le Jeu* was penned, Marozzo includes an identical technique for the *ronca*, or Italian bill:

> [W]hen he throws a cut to your head with his bill, you'll defend yourself by beating it forcefully inwards with the heel of your bill, to your right. In that beat, hit him in the face with the heel of your bill, fixed on your left foot though advancing it just a bit.
> (Marozzo, 2018, p. 287)

This ubiquitous play points us toward three important features of fencing with heavy polearms: finesse, the liberal use of the butt end, and the importance of thrusts.

We have already looked at the importance of using finesse to protect the physical integrity of the weapon. We should not overstate the fragility of these weapons, however. As mentioned before, fighting polearms could be reinforced with langets (though we should hasten to add that not all were—langets do add weight to the weapon, after all, and can adversely affect its balance), and I do not wish to be understood as saying that polearms snapped like twigs every time they were swung strongly together. The larger point to be made about polearm finesse is that which is hinted at by the Anonimo Bolognese: the risk of overswinging is very real. This is not to say that titanic polearm swings are never a viable attack; we'll discuss them shortly. It *is* to say that a heavy polearm

fencer must carefully choose the moment for such attacks lest they leave the fencer open to a more conservative opponent. *Most* polearm fencing, even with heavy polearms, is conducted with fairly small, subtle, and easy-to-redirect movements not unlike those of a spear.

The use of the butt end dovetails with this need for finesse. In his description of the poleaxe, Monte notes specifically that a poleaxe should have a strong spike at the butt, because the butt of the poleaxe it is often used for fighting.

As a practical matter, because most polearms are head-weighted, it is often faster to use the head of the weapon not for striking but as a counterweight to the lighter butt end (poleaxe plays are particularly notorious for this). When an attack with the head of the weapon is parried, the butt end can frequently rotate around the parrying weapon to strike the opponent (unlike in spear fencing, this technique is used quite liberally with heavy polearms). This same basic maneuver can be used to initiate many grappling plays. Lastly, the butt end of the weapon can often thrust and harass more quickly than can the head. Talhoffer demonstrates this in one of his treatises, showing the axe held overhead, point forward, with the caption "here he will distress with stabbing and hitting" (Talhoffer, 2019). *Le Jeu* refers to this butt-forward stance as a mark of a highly skilled axe-fighter.

This brings us to the question of thrusts. When I was young, I assumed that heavy polearms were primarily used to deliver heavy cuts or percussive blows. Why else would they have such wicked-looking heads?

It may be that historical fencers labored under similar misconceptions, for it is not hard to find in our extant heavy polearm treatises reminders to make frequent use of the thrust with head or butt. *Le Jeu* ends with an admonition not to forget to harass an opponent with thrusts to the face and feet. The Anonimo Bolognese includes among its general instructions for poleaxe fencing, "Know that you must never refrain from striking your enemy with a thrust to the face or some other part of his body that is unarmored, like the testicles or the groin" (Anonimo, 2020, p. 236). Manciolino goes even further. Even with the bill, he teaches no attacks but thrusts, writing "let me end the book by repeating what I have already said. With all polearms, there is only a good way of striking, which is with the thrust" (Manciolino, 2010, p. 145).

The thrust performs several vital functions that a cut or percussive blow does not, even with a heavy polearm. For one thing, it gives a fencer techniques to use with a heavy polearm that function in a close-order formation—an important consideration for the medieval or early Renaissance student, whose only "military" weapons training would be that which he paid for privately. The thrust is also the easiest attack with which to feint and redirect, because it allows the fencer to make use of the length of the haft to redirect the head with small movements of the back hand (for more on this, see Chapter 10 on spears). It is also the easiest attack with which to target the extremities of an opponent's body. Consider, for example, that a polearm-length weapon can target the foot fairly easily with a thrust (remembering especially that the butt ends of many heavy polearms were spiked); doing so with a swinging blow is fairly awkward, and easily defended simply by pulling the foot back, out of the arc of the incoming blow. These two features of the thrust make it ideal to harass and discomfit the opponent, as *Le Jeu* recommends. Further, as the Anonimo Bolognese' advice suggests, the thrust is generally the fastest attack a heavy polearm can make, and it is therefore most likely to be the attack with which a fencer can exploit momentary windows of opportunity. A heavy polearm fencer who neglects the thrust can easily spend the fight constantly reacting to the faster attacks of one who remembers it. Lastly, as we will discuss in more detail in Chapter 12, the ability to plant a point in an armored opponent's mail in order to inflict blood loss, control movement, and ultimately throw the opponent to the ground is a critical part of historical armored combat.

Let us examine a play that exemplifies these characteristics of the thrust. From Manciolino's discussion of fencing with the bill:

> With a smart right-foot *accrescimento*, push a thrust to the opponent's face—then push the iron downward and do a *straziamento* to his arms, and push a second thrust to his chest. Retreat by means of a rearward jump, and set yourself a left-foot-forward guard with the iron high in the air.
>
> (Manciolino, 2010, p. 143)

Here, the attacking fencer feints a thrust at the opponent's face, but the fencer's true aim is to use the bill's sharpened hook to lacerate the opponent's forward arm with a dragging motion (the *straziamento*). The feint is a simple one, requiring the attacker only to push the weapon down a bit to turn what appeared to be a thrust into an arm hook. With the opponent thus off balance and injured, the attacker can perform a quick, more lethal follow-up by slamming the bill's point into the opponent's chest. To retreat, the attacker reverses the weapon so that the butt sweeps up, parrying any follow-up blows and threatening either a quick jab with the butt or a deadly blow with the head should the opponent not be incapacitated. The play is quick, subtle, and makes good use of all the features of a heavy polearm. It also teaches skills—feinted thrusts and raking slices or off-balancing arm drags—that are functional even in close order. It is worth noting that even if the opponent is armored, or the attacking fencer is armed with a poleaxe or other weapon with an unsharpened hook or pick, the arm drag in this play can still perform its essential unbalancing function, and the thrust can easily be redirected to the opponent's more vulnerable throat instead of a breastplate.

Sidebar: How many of these things are there?

Europe produced a staggering array of ironmongery on sticks during the Middle Ages and Renaissance. This enormous variety has long fascinated a certain sort of weapons nerd.

The various names modern people use for polearms are, like most names modern people use when we classify swords, mostly a backward-looking projection onto the past. Perhaps it is helpful to differentiate between a halberd and a voulge if one is a museum curator, but from a fencer's perspective, I find attempts to invent a single naming convention for every single polearm Europe ever produced to be singularly unhelpful. Indeed, as a fencer, the great variety among polearm irons matters surprisingly little. At the end of his section on bill fencing, Marozzo writes:

> Understand that these matches with the bill can also be done with the halberd or with the pollaxe. For my part, I make few distinctions among them … and consider one manner of play to apply to all three of these arms, namely the bill, halberd, and pollaxe.
>
> (Marozzo, 2018, p. 288)

And yet, the question of why *this* blade looks a certain way and *that* blade has a hook is still a fair one.

Here's how I tend to evaluate European polearms from a fencing perspective. Rather than trying to memorize two dozen individual names, I tend to think of polearms along these dimensions:

- Sword-like blades are good for stabbing and cutting.
- Axe-like blades are good for cutting, and better than sword-like blades.
- Spikes are good for stabbing, and better than sword-like blades.
- Hammers are good for bashing and not being deflected by armor.
- Picks are good for hooking and bashing, and good at concentrating force.

- Bill hooks (sharpened) are good for hooking and cutting.
- Prong hooks (unsharpened) are good for hooking, and better than bill hooks.

As a rough rule of thumb, when "designing" a polearm (or, more realistically, when commissioning one), you can pick up to three of these features. For each additional option you pick, the whole thing becomes more expensive and all of your chosen options get slightly worse, but you may have added capabilities you didn't have before. Choose your options wisely!

Then decide length. Short polearms (~6 feet) have no minimum range and are fast, and not good at mutual support in a formation. Long polearms (~9 feet) have a minimum range and are slower, but hit harder and are good at mutual support in a formation.

And that's basically the game.

It's worth emphasizing that the differences in capability here are, in my experience, less significant than non-fencers frequently assume. From a purchaser's standpoint, they might well be significant—anybody purchasing a polearm would want to ensure that he is spending money on a weapon that fits his individual style, after all. From a fencing teacher's perspective, though, the differences are very small. These are all extremely powerful weapons, so the difference in, say, cutting capacity between a sword-like blade and an axe-like blade is not always material when it comes to how one should use the weapon. In addition, many of the "specialist" capabilities, such as those afforded by the various form of hook, are not the sort of thing that a fencer can take advantage of on a whim. The core of fencing with all polearms is essentially the same, regardless of the shape of their irons. No wonder, then, that period fencing treatises tend not to treat the great multiplicity of European polearms individually.

If this was where our discussion ended, my teenaged self would be very disappointed. Did people never deliver heavy blows with the great iron heads of these weapons?

Fortunately for my teenaged self, they did. And quite powerful blows they could be. A single example from Fiore should suffice. Describing a poleaxe strike to the helmet using the hammer face, he says, "you'll likely drop to the ground dead after being struck in the head like this" (Fiore, 2017, p. 37r).

The difficulty with using the head of a heavy polearm for swinging blows is that such attacks are also comparatively ponderous (speaking, as always, compared to other weapons—in absolute terms, the swing of a heavy polearm can still be completed in under a second). Monte notes, "For the hammer or the blunt part, is altogether potent in matters of strength for feinting or striking, but for fighting the sharp points are best" (Monte, 2018). In other words, while the hammer face of a poleaxe is useful for feinting, earnest poleaxe fencing relies mostly on the sharp spike at the head and butt of the weapon.

Yet, as Monte points out, the mere fear of a blow from the head of a heavy polearm can serve a purpose. This is one reason why the head is so useful for feinting: such attacks can be so devastating that no opponent, no matter how heavily armored, can afford to ignore them. Thus the following play from Marozzo:

> [F]ind him with a feint of a fendente to his head in order to make him parry while you step straight toward him with your right leg. Then, as he parries the blow to his head, you'll draw your bill to yourself a bit, and once you've done so, throw a thrust to his chest.
> (Marozzo, 2018, p. 287)

Here, the fencer feints a descending cut to draw the opponent's weapon high to parry the fearsome overhead blow of the bill. As soon as the opponent does so, the attacking fencer can quickly pull the descending weapon back, avoiding the parry entirely, and thrust home. Critical to this calculus, of course, is the fact that sometimes the swinging blow of a heavy polearm *is* real. Marozzo's deception is not a complicated one, but it is a bold fencer indeed who does not move to parry an overhead cut from a bill.

The great challenge in employing the head is in obviating the timing difficulties inherent in their ponderous swings. The first way to do this is through careful understanding of the timing of weapons. Marozzo gives an example, the fencer beginning with the head of the bill pointed low and diagonally forward to his left.

> Then stand fast, and if he attacks you again, defend yourself by beating his blow upwards and then attack downward with the beak of your bill, either diagonally or straight.
> (Marozzo, 2018, p. 288)

As his opponent attacks, the fencer parries by lifting the bill up and to the right, beating the opponent's weapon to the defender's right. From here, the defender can slash back down with a mighty cut before the opponent's own bill can recover to defend.

Another possible way to achieve the window of opportunity necessary to strike a swinging blow was to make the opponent temporarily lose grip on the weapon. We have already seen this briefly in Chapter 10, discussing how strong spear parries can knock a spear out of an opponent's hand. The same principle receives even more play with heavy polearms.

Le Jeu accomplishes this when the polearms are crossed (what *Le Jeu* calls the *demy hache*, or half-axe) by forcefully running the butt end of the axe down the shaft of the opponent's weapons to dislodge the hand. As the opponent cannot easily manipulate a heavy polearm with one hand, this creates a window with which to strike a heavy swinging blow.

Lastly, of course, is the possibility of creating an opportunity to strike through grappling. For instance, Fiore makes his "you will likely drop to the ground dead" comment after disarming his opponent of his axe by twisting it out of his hands; he has the opportunity for a titanic hammer blow because his opponent no longer has a weapon. In another armizare play, the attacker places the opponent in an arm lock, choking up on the axe with one hand to deliver a blow.

Heavy polearms are themselves quite useful grappling aids, a fact that likely explains their popularity against armored opponents at least as much as the force of their attacks. Grappling techniques are especially prevalent in poleaxe fencing, but their adaptation to other heavy polearms is usually straightforward.

"Grappling," in the polearm sense, does not necessarily require fencers using their hands. In the sense I am using that term here, grappling techniques include any use of the weapon to help trip or unbalance the opponent. We have already seen one such technique, from Manciolino: heavy polearms almost always have a pick beak, bill hook, or other projection that can be used to drag an unwary opponent off balance by the arm (in a pinch, even an axe blade can be used this way). Another significant use of such hooking techniques is to catch the opponent's weapon itself, dragging it out of the opponent's hands or unbalancing an opponent who attempts to retain the weapon (which could, of course, provide the opportunity for a mighty swinging blow).

Another popular grappling technique was to hook the opponent's leg with the polearm head and yank the opponent off balance. Such an attack could come from a swinging blow, as in this play from Paulus Kal:

If he strikes from above, then parry with the lower part of the axe held above. Or let fly at him with a sweep and hook the axe inside the right knee to pull him down.

(Kal, 2019)

The opportunity can arise from a low thrust, as well. As the Anonimo Bolognese reminds its readers, the canny polearm fencer must always be ready to hook the opponent however and whenever the opportunity may arise:

You are also advised to throw your enemy to the ground every time you are scuffling with him, by hooking his legs, arms, neck or some other parts of his body with the horn of your poleaxe.

(Anonimo, 2020, p. 236)

Despite the prevalence of hooks and horns that could catch an opponent, however, the most popular way for a heavy polearm fencer to grapple was to cast the butt of the weapon across the opponent's neck (or, in some cases, to place the butt under the opponent's chin). This is especially effective against armored opponents, as the need to armor the head and chest most heavily shifts an armored opponent's center of

balance upwards. Variations on such techniques are numerous. *Le Jeu* includes one in which the defending fencer catches an incoming overhead blow with the axe held horizontally overhead, on the haft. The defender then steps forward with the left foot, placing it behind the opponent's heel, shoves the axe up (otherwise, the attacking fencer could imbalance the defender with a pull of the axe) and rotates the axe so that the butt is under the attacker's chin. A hard thrust to the chin, and the opponent falls to the ground.

Heavy polearm fencing, and poleaxe fencing especially, is particularly concerned with throwing the opponent to the ground (11 of the Anonimo Bolognese's 13 poleaxe plays end with throwing the opponent to the ground—four of them with a thrust to the groin or testicles). This reflects an important assumption that underlies much of the fencing behind these weapons: that they were used by and against fencers in plate armor. *Every* surviving poleaxe treatise—whether armizare, Bolognese, KDF, or otherwise—makes this assumption explicitly. It is worth saying a few words about the implications that this does or does not have when imagining the use of a heavy polearm without armor, based on modern sparring experience.

When fencing against an unarmored opponent, the essential elements of heavy polearm technique are finesse, the canny use of both ends of the weapon, and a reliance on thrusts until and unless an opportunity arises for a heavy swinging blow. These are the same essential elements that we discussed at the beginning of this chapter; they form the foundation of heavy polearm fencing both in and out of armor. What *is* different when using a heavy polearm against an unarmored opponent is that hooking, tripping, and other grappling techniques occur far less often. Such techniques are no less *valid* against an unarmored foe; it is simply that the fight tends to end before the need for such techniques arises (a notable exception is techniques used to hook the opponent's weapon itself, the occasion for which arises every bit as often in unarmored polearm fencing as in armored). While (comparatively) ponderous in the cut, heavy polearms are far too fast and subtle in the thrust for the fight to end in grappling with any regularity absent the protection of heavy armor.

Indeed, without armor, fencing with heavy polearms is essentially the same murderous lunacy as fencing unarmored with spears. It is an excellent way to come face to face with the power and speed of these weapons—as well as to understand why so many treatises assumed that nobody *would* use such weapons without armor.

And with that being said, it is time at last to discuss in greater detail just how to fence against various kinds of medieval armor, with various kinds of weapons.

Which would win: halberd or greatsword?

Both the halberd (or bill, or glaive, or even poleaxe; the exact type of heavy polearm does not matter for present purposes) and the greatsword are large, hard-hitting weapons. Both have a history as specialist weapons in the nascent professional armies of the Renaissance, and both have been used as parade or ceremonial weapons for high-status individuals. But how do they compare to each other?

Achille Marozzo discusses this question directly, and he gives the advantage to the polearm. Here is the beginning of his discussion on the subject:

> This is a conflict between someone who has a two-handed sword and someone else who has a polearm of whatever kind ... I will show you the method and the way in which you can valiantly and safely defend yourself and present grave danger to your enemy so that he can do nothing to harm you, almost without fail, if you have a bold heart. However, I advise you against making such a match, because the advantage will always be his, in my opinion.
>
> (Marozzo, 2018, p. 264)

Marozzo proceeds to discuss two scenarios: one against a polearm fencer who is not trained in fencing, and one against a polearm fencer who is. In the latter case, against the skilled opponent, he recommends an unorthodox reverse grip, with the right hand gripping the ricasso of the sword behind its parrying lugs and the knuckles of both hands pointing up, with the sword point toward the swordsman's right side. This grip shortens the swordsman's reach but provides the superior speed and leverage necessary to oppose a polearm in skilled hands. He warns, however, that it will not work against a bill or other polearm with hooks or projections, because the forward hand would be at risk against such weapons:

> This method that I will place here for you ... will be a very useful thing ... no matter which [pole] weapon they may have, unless it's a bill or a spiedo ... because of the concern that you should have for your right hand, which would be in danger from the horns of the spiedo or the forward beak of the bill.
>
> (Marozzo, 2018, p. 265)

In fact, Marozzo provides *no* remedy for a two-handed swordsman against a skilled opponent armed with a bill (or halberd, or poleaxe, as he himself states that he considers the three weapons essentially interchangeable), and his silence speaks loudly.

There are two essential problems for the greatsword that Marozzo identifies: reach and speed. Large as a greatsword is, it is still only the height of a man (or, recalling our discussion in Chapter 8 of the Bolognese two-handed sword, slightly less than the height of a man). Most polearms are taller than that, and virtually none are shorter. As a result of this reach disparity, the initiative will almost always lie with the polearm fencer, and Marozzo strongly advises the greatsword fencer not to be the first to press the attack. Doing so (as I can confirm from my own experience on both ends of this matchup) is an invitation to be stabbed in the face as the greatsword fencer attempts to cross the distance necessary to bring the weapon into play.

Yet close the distance the greatsword must. The sword fencer's only reasonable hope of doing so is to bait the polearm fencer into attacking first, so that the polearm attack can be parried as the greatsword rushes in. This is itself a perilous proposition, and Marozzo's discussion of this uneven matchup is replete with warnings not to underestimate the speed with which a polearm can attack, pull back from a parry, and attack again. A bold greatsword fencer with quick reflexes can do it, but the advantage remains firmly with the polearm fencer.

If the polearm fencer "doesn't understand the use of arms," as Marozzo puts it, the greatsword fencer can follow a simple program of sweeping aside the incoming thrust (recall that even with a heavy polearm, thrusts will predominate) and rushing forward, keeping the sword on the opponent's weapon so as to maintain control of it, and then unleashing an attack (Marozzo, 2018, p. 264). The execution is difficult, but the concept is simple.

Against a skilled polearm fencer, however, the matchup becomes far more uneven. Attacks will be faster and stronger, to the point that Marozzo recommends his unorthodox grip and far more conservative footwork to compensate, and reiterates his advice to the greatswordsman to "be vigilant" (Marozzo, 2018, p. 266).

As discussed in Chapter 8, the greatsword, while large, is still fundamentally balanced like a sword. It is *not* essentially a sword-shaped polearm, as I have sometimes heard it described. Nowhere is this more evident than when it faces an actual polearm. When held with both hands on the handle, a greatsword simply is not fast enough to reliably respond to the rapid-fire thrusts of a skilled polearm fencer. Marozzo's grip with a hand on the blade gives the greatsword a more polearm-like grip on the weapon, but this is only a partial remedy.

A skilled polearm fencer, following the rules we have discussed in this chapter, will be unlikely to overcommit to a thrust so much that the polearm can be easily parried and closed with. A skilled polearm fencer who is closed with will be much more likely to bring the other end of the weapon quickly into play, and perfectly comfortable using a heavy polearm even at grappling distance. The skilled fencer will also be ready and able to use the various features of the polearm's iron to dissuade the sword fencer's from adopting Marozzo's polearm-like grip—and, as Marozzo warns, will likely succeed.

All of this gives the heavy polearm a bigger advantage against a greatsword than it might first appear. And the situation is even worse if the combatants are armored, for while a greatsword is far from impotent against armor, it has nothing like the fearsome anti-armor capabilities of a heavy polearm. As Marozzo says, this is a matchup that the greatswordsman is best advised to avoid.

CHAPTER TWELVE

Taking hits: fencing in, and against, armor

> *On the other hand, if you fight at the barrier and are well armored, you can still win the fight even if you take a lot of hits.*
> —Fiore dei Liberi (Fiore, 2017, p. 1v)

So far, we have only alluded to the effect of armor on fencing techniques. We have seen how certain weapons lend themselves more than others to armored combat, but a full discussion of the use of medieval weaponry against medieval armor has been deferred until now.

Discussions of this sort often fall into one of two extremes, both of which I will strive to avoid in this chapter. The first extreme is the belief that armor did little or nothing to protect its wearer, as if it were nothing more than a fancy costume. Examples of this belief are not hard to find in film and television. Warriors on-screen are forever cleaving their armored opponents in twain or running them through, with no apparent resistance from their armor. If this were the historical reality, we might well wonder why anybody bothered to wear armor at all!

But the second extreme is just as bad: the belief that medieval armor made its wearer invincible. If this were the case, we might expect medieval combat to have been an orgy of slaughter, invincible knights mowing down lesser-armed peasants like grass. Often this misconception is paired with another misconception, that armored knights were finally and abruptly made obsolete by the invention of the gun. Both of these are false as well (in fact, plate armor coexisted on the battlefield with shoulder-fired long guns and pistols for many centuries).

How, then, *did* medieval weapons—and especially bladed weapons—work when used against armored opponents?

The first thing to say about fighting armor is, of course, that not all armored opponents were armored from head to toe. The easiest way to injure an opponent in partial armor is to strike wherever the opponent's armor does not cover. This is only common sense, but it raises the question of how hard it really is to make a "called shot" to an opponent's unarmored openings.

The answer to that question is complicated. It is not *inherently* difficult to strike around an opponent's armor. However, it can become *very* difficult to do so while defending against an active, resisting opponent. On the one hand, a skilled armored fencer should be able to strike a target as small as the sights (eye slits) of a helmet from a spear's length away, so striking around partial armor poses no particular challenge from a purely mechanical standpoint. On the other hand, even partial armor narrows a fencer's "weak spots" to a few, highly predictable areas. Knowing that an opponent is likely to target one's unarmored openings is half the battle. Essentially, even partial armor gives a knowledgeable fencer foreknowledge of an opponent's likely attacks, which makes them significantly easier to defeat.

As a result, a fencer could need to have recourse to anti-armor fencing techniques whether facing an opponent in full or partial armor. Before we discuss what those anti-armor techniques are, we first need to discuss the defensive properties of the armor types they were intended to defeat. We will discuss three types of armor a sword in the Middle Ages had to contend with: plate, mail, and linen.

By "plate," I mean shaped sheets of metal. The sheets can be large or small, thick or thin; for our purposes it doesn't matter.

Plate offers four defenses to the wearer:

1. Its shape deflects blows, making them less likely to land solidly.
2. Its material makes it difficult to penetrate with edges and points.
3. Its size diffuses the impact of a blow over a larger area than the striking surface of the weapon.
4. Its mass absorbs the momentum of blows.

By "mail," I mean links of wire or rings of metal that interlock to form a sort of metal fabric. There are many patterns in which one can "weave" mail, and a lot of subtle variation—how thin or thick the individual links are, how large or small the interior diameter of the rings is, whether the links are made of drawn wire or punched out of sheet metal—that depend upon time and place. For our purposes, again, it doesn't really matter. Mail offers two defenses:

1. Its material makes it difficult to penetrate with edges and points.
2. Its mass absorbs the momentum of blows.

Linen is an important but often overlooked material in the history of body armor. This is not to say that all linen garments, or even all linen garments worn with armor, functioned as armor in their own right. A jacket made of 20 or even 30 individual layers of linen is armor. A linen arming doublet made of two or three layers, worn under plate armor, is not.

Linen offers up to two defenses:

1. Its material is resistant to cutting.
2. If a linen garment is made of many individual layers, its construction absorbs the momentum of blows.

At this point, my teenaged self would interject with a question: what about leather armor? Leather armor is a staple of fantasy armor, frequently treated as a light option for poor or very agile fighters.

Leather is a vital component of many types of armor. It's used to make sturdy, flexible, and weather-resistant straps, laces, and foundations for rivets. However, it is almost entirely absent from extant treatises as a protective material. Monte does recommend using a leather chest reinforcement when traveling on a road by night, seemingly in place of or supplementary to mail hidden under a traveler's clothes, but otherwise, leather is treated as essentially no defense when it is mentioned at all. Interestingly, Monte says that the leather reinforcer he recommends should be worn somewhat loose about the torso so that it absorbs a weapon's impact, which suggests that he, too, did not place much stock in leather's inherent ability as a material to resist attacks.

Knowledgeable readers may yet object that, while ordinary tanned leather is easily penetrated by blades (it is only flesh, after all, at the end of the day), leather can be made into a very hard protective material by "boiling." It is certainly true that "boiled leather" is a term that appears in armory inventories, but the exact nature of boiled leather armor remains difficult to pin down. There may not ever have been a uniform meaning to "boiling" leather; it is not hard to imagine individual leatherworkers keeping their particular process and ingredients a closely guarded secret. This makes it difficult to conduct modern tests on boiled leather armor using sharp weapons. Some boiled leather preparations are demolished in test cutting with frightening ease, while others put up more resistance—but how are we to know which preparations represent historical boiled leather?

Fortunately for us, medieval fencing treatises focused on metal body armor, which was presumably sturdier than any boiled leather formula in any case. And so, in keeping with our practice throughout this book, we shall leave further discussion of boiled leather to experimental archaeologists and follow the treatises where they lead.

Armor evolved very rapidly in the Late Middle Ages. Even when armorers were not pushing the envelope, medieval armor was the subject of a dizzying array of variation according to regional style, intended use, and personal preference (Monte, for instance, has sharp criticism for those who prefer armor that he deems too thick). It may be surprising, therefore, to hear that most of these variations are irrelevant for the purposes of this chapter. They were certainly of great interest to the makers, buyers, and wearers of armor. How could they not matter to the fencing masters who taught how to use and defeat armor?

The answer has to do with the difference between individual coaching and designing a martial arts system or curriculum. A fencing master would, presumably, work with an individual student to maximize the benefits of that student's particular suit of armor (or "harness," as historical fencers call it). However, a martial arts system's body of anti-armor techniques needs to be broad enough to accommodate the wide variety of armors that its practitioners might encounter. Thus, the effects of most historical armor variation can be ignored as we seek to understand the breadth of medieval armored fencing techniques.

Now that we've discussed how armor works, what about bladed weapons?

For the purposes of this chapter, weapons have essentially four ways of inflicting injury, which I shall refer to as hew, thrust, slice, and smash.

As these are all modern terms, and not entirely unambiguous even in the historical fencing community, let us define each of these terms before moving forward.

A hew is a type of cut that operates like a guillotine or a meat cleaver: the edge of the sword impacts the target at high enough speed to open a crack in it, which is then propagated through sheer momentum; the sword blade runs through the target like a pair of scissors being run down a roll of wrapping paper. Straight-bladed swords, such as the ones we have discussed in this book, cut most deeply when hewing.

A thrust is a stab: the point of the sword impacts the target with enough energy to pierce a hole in it, which becomes a wound channel as the blade is pushed deeper into the target through that hole. It will be important to understand the discussion that follows to know that the shape of a blade can affect a thrust. Blades with sharp edges cut their way through the target (essentially slicing it from the inside; see below); blades without sharp edges perform a minor amount of crushing the target's tissues as the thicker parts of the blade force their way inside. Also critical for the discussion that follows is going to be this: while hews travel through the target on momentum alone, a fencer can continue to apply force to a thrusting sword blade.

A slice is a type of cut that works like a saw or steak knife: the edge of the blade is drawn or pushed along the target's surface. This tears open a crack on the target, which the blade can propagate as it continues to be drawn or pushed along and into the target. This is the optimal way to cut with many curved swords (though there's a conversation to be had about mixing hewing and slicing mechanics depending on exactly how curved the sword is).

By "smash," I mean inflicting blunt force trauma to the target. There are a number of ways to do this, which we'll get into shortly, but they all follow the same basic wounding principle: transfer as much energy in as short a period of time to the target as possible.

We have now looked at how different types of armor defend, and the different ways in which weapons can wound. At last we are ready to see how these two things interact.

We shall begin with linen, the lightest form of medieval armor. As noted above, it's important to note that a few layers of linen, such as the arming clothes worn under plate armor, don't really count as armor. Arming clothes aren't even necessarily padding (as a general rule, the amount of padding in arming clothes decreases over time, as the percentage of plate coverage in plate armor increases); they're just a suspension and anti-chafing system. But people did make actual armor out of linen in the Middle Ages and Renaissance, by simply making garments out of 20 or 30 layers of linen.

Such "padded jacks" were essentially immune to sword cuts. Linen fabric is naturally cut-resistant, and the layers of a jack are not tightly compressed. This means that each layer can give slightly before a cut, like a curtain. The cumulative effect of these cut-resistant curtains is to rob a cut of its cutting capacity and its percussive impact. Even under experimental conditions, the best modern longsword fencers cannot reliably get through 30 layers of historically accurate linen. Linen armor would still get cut up in combat, of course, but it would keep the edge out of a fencer's skin and dampen the force of hewing or smashing blows. Slicing linen is easier than hewing, but the damage of a slice is mitigated by the sheer thickness of multiple layers of linen. Combined with mail, linen could form a very tough composite armor system (at the expense of swaddling the fencer in something like a hot and heavy moving blanket).

How could such an armor system be defeated? The critical consideration is the fact that cuts are based on momentum rather than continual application of force. As linen (or linen and mail) armor defends primarily by continually robbing blows of force, a successful attack must have so much force that the armor can't sap it all, or be of such a nature that force is being continuously applied. Polearm strikes do the first; woe betide the fencer who trusts a padded jack to nullify the force of a poleaxe, bill, or halberd cut. Thrusts do the second: a sword may not be able to reliably cut through linen armor, but it can certainly stab through it.

One interesting note about thrusting through linen: it is extremely beneficial for the thrusting weapon to have an edge. What's really going on is that the point finds a spot between linen threads, and the forward motion of the weapon forces threads along the sides of that point. If the point is supported by a cutting edge, such as with a spear or a sword, the threads are sliced open as they are forced along the edge. An edgeless spike, such as an estoc (essentially a long, edgeless spike on a longsword hilt) or some rondel daggers, will merely crowd the threads together. This isn't nearly as efficient, and may even fail to penetrate altogether if there are enough layers. Conveniently for armor-wearers, mail can protect linen from being cut by a sharp-edged thrust, while linen can protect mail from being burst by a slim, edgeless spike: another way that linen and mail complement each other as a composite armor system.

Let us now discuss metal body armor. The first concept to remember when discussing historical anti-armor fencing is that even an opponent who appears to be armored from head to toe can't *actually* be wearing armor everywhere. This is distinct from the scenario of fencing against partial armor. When fencing against partial armor, it is easy to see where an opponent is unarmored. A skilled armored fencer, however, can see even the non-obvious gaps in the opponent's protection. KDF refers to these gaps as "openings." Liechtenauer phrases it thusly in the zettel, glossed here by Pseudo-Peter von Danzig:

> *Leather and gauntlets, below the eyes, seek the openings correctly.* Gloss—*Note*: this tells where the best places are to attack an armoured man through the harness. These are under the face, under the armpits, in the palms of the hands, on the arms from behind into the gauntlets, into the hollows of the knee, or below on the soles of the feet, in the insides of the elbows, between the legs, and anywhere the harness has its articulations.
> (Von Danzig, 2010, p. 148)

Ringeck adds that "at these locations one is best able to succeed, that is why you shall know how to seek the openings correctly" (Ringeck, 2019). The lesson here is simple in concept but unintuitive in application: no harness covers a fencer completely, and a successful armored fencer must learn to see the gaps even when they are not immediately apparent.

As the gloss says, the palm of the hand is almost always unarmored, so an armored fencer may be stabbed in the palm even when wearing gauntlets. Just as the palm is usually unarmored, the visor is often kept up in melee combat, to permit greater situational awareness and easier breathing. Hence, an armored fencer may be stabbed in the face. If the visor is down, the attacker can close to grapple, lift the visor up, and *then* stab the opponent in the face (Fiore presents this attack twice, once with a longsword and once with a poleaxe).

226   THE USE OF MEDIEVAL WEAPONRY

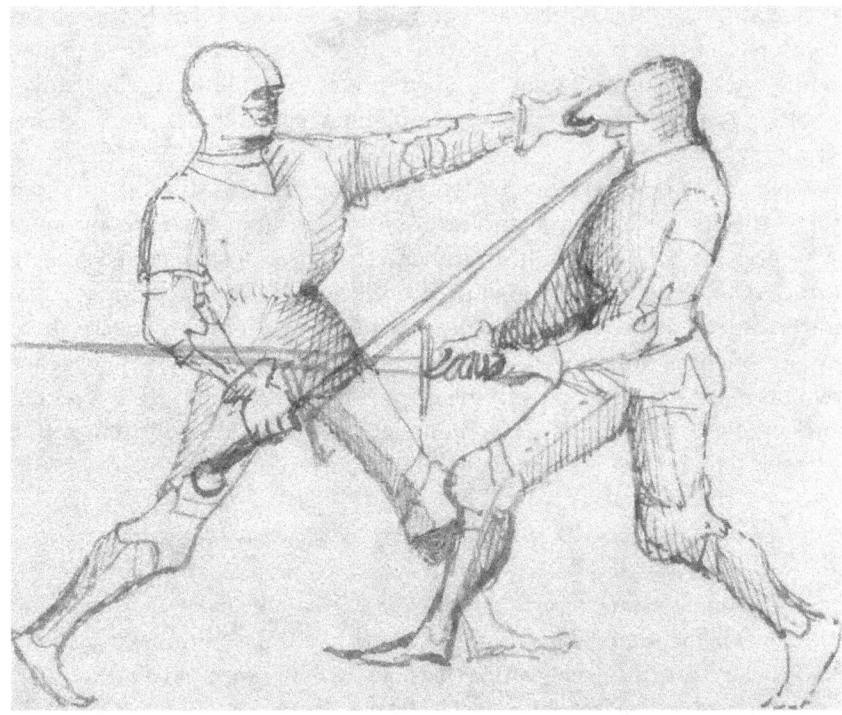

Additional potential openings, even in "full plate armor," include the back of the knee, the buttocks, straight up between the legs, the back of the hand up the gauntlet, and anywhere else a harness of plate armor is articulated to permit greater freedom of movement (particularly the inner elbow and armpits).

Sidebar: What about armored palms?

When I talk to people about gauntlets, I often find that they are disappointed to find that the palms and undersides of the fingers were typically unarmored. I think that there is something compelling about the image of a fragile human hand encased in metal that we find romantic, and it seems anticlimactic to find that the mighty steel fist has a soft underbelly.

This disappointment is frequently followed by a question: couldn't you protect the underside of the gauntlet with mail?

The answer to this question is literally yes, but it comes with some important caveats. Monte *does* advise protecting the left palm with strips of mail, noting that this will allow the left hand to seize an opponent's blade more safely when the time comes. However, he recommends these strips of mail not as protection against being stabbed in the palm (the point may not strike one of the mail strips, after all), but rather as protection against being *sliced* in the palm when grabbing the opponent's sword by the blade.

Why not, then, protect the entirety of the palm with fine mail? Such a gauntlet would certainly make the palm better protected, but it would also result in significant loss of feeling and control in the left hand. Indeed, a loss of feeling is noticeable with even a few strips of mail. Since the hand's role in a fight is first and foremost to control a fencer's weapon—and bearing in mind the importance of precise targeting in armored combat—it is not hard to infer why Monte's advice was not universally followed.

Many of these openings were covered with mail, but not all of them, and not all the time. Medieval armor was highly configurable. Some fencers preferred to leave the mail off of some areas, to reduce the weight of their armor or to optimize their armor for mounted combat (it is possible to control a horse with fully enclosed legs, but it is hardly optimal). A skillful and perceptive fencer was expected to be able to take advantage of these vulnerabilities. By contrast, Monte notes disdainfully that a truly skilled fencer can be almost entirely unarmored from the rear and still be perfectly safe: "He who knows how to govern himself in the way we have already prescribed can be very easily protected, to keep him from being caught by the back part, or even by the sides" (Monte, 2018).

In many cases, of course, these openings *were* defended by mail, which brings us to the question of how mail armor was defeated. What follows is given to us in the context of our manuscripts as advice for defeating plate-armored individuals through their mail, but it is generally applicable to fencers who are armored only in mail, as well.

When I was young, I labored under the misapprehension that mail was a "light" armor. This is incorrect, or at least misleading. Mail was, after all, the gold standard in metal body armor in Europe for nearly a thousand years, and provides excellent protection against most weapons.

Mail is especially effective against cuts. People can, and do, cut through mail under experimental conditions, but in combat conditions, it is effectively impossible. This always raises the question of whether or not a fencer can simply bludgeon an opponent into submission through mail. After all, mail is (relatively) flexible. Is it a viable strategy, when attacking mail, simply to swing really, really hard?

The answer to this question is both yes and no. On the no side, it is important to remember that mail may be flexible, but it is not *infinitely* flexible. It cannot be folded up like a thin piece of silk, for instance; the individual links don't bend. The mail's relative inflexibility still spreads out the force of a blow somewhat. Mail is also heavy. That's part of how it does its work. The sheer mass of metal a weapon has to move helps to absorb the force of a blow. Mail is also not worn alone, over bare skin. Mail is always backed at least by a few layers of fabric. There may be actual padding behind it. There may be padding *over* it, as well. All of that adds to the mass a weapon has to displace, as well as possibly robbing the cut of force through the curtain effect of textile armor.

As a result of all this, simply swinging away at mail with a sword, or even an axe or bludgeon, is not an optimal strategy from a wounding standpoint. This is not to say, of course, that blunt impact through mail is completely ineffective. Swinging really hard at an opponent's mail can injure a fencer, and it is certainly better than just giving up.

It is not, however, better than thrusting.

Thrusting with a bladed weapon into mail is a topic that is frequently misunderstood. Part of this misunderstanding is related to understanding how different medieval weapons thrust. A heavy polearm with an armor-piercing spike, such as a poleaxe, can absolutely go through straight through even well-made riveted mail. Swords don't do that. However, they *can* inflict penetrating wounds through mail.

One of the big differences between a metal "fabric" like mail and actual fabric is the strength of the individual "threads." If a sword point strikes the thread of an actual textile, rather than slipping between the many threads that make up the fabric, it can sever that thread and continue to penetrate the fabric. If a sword point strikes a mail ring, though, it does not penetrate. A sword point can't be pushed through the metal of the links themselves.

Consequently, when thrusting at mail with a sword, it's important to get a good "plant." The sword point must actually slip *into* the rings. The pointier the sword, the more easily this will happen, but it isn't necessarily easy against a moving target (not coincidentally, European swords became sharply pointed right around the rise of plate armor, when mail became an even harder target). Sword thrusts don't always find purchase.

Of course, because most sword blades get thicker toward the hand, even if the point is firmly planted in a ring, at some point that ring will stop the blade from pushing in any further. At this point, we return to that most armor-relevant feature of the thrust as compared to the cut: a fencer can continue to put force behind a thrust. Because thrusts are not primarily momentum-based attacks, a fencer can continue to shove the sword forward even after it's been arrested by the mail link.

Under strong pressure against a resistive material (like a mail link that doesn't want to be burst), a long and thin piece of steel (like a sword blade) will want to bend. This is one reason why swords being thrust into mail tend to be gripped on the blade: the forward hand acts as a stabilization point so that the part of the blade going into the mail isn't as long, and hence not as likely to bend, and hence more likely to be shoved deeper into the opponent's body. This practice of gripping a sword blade with the off-hand is most famously associated with longswords, but it can be done with virtually any straight medieval sword. Different traditions have different names for it: "half-swording" is used by some German authors, such as Talhoffer; other German authors tend to prefer the terms "short sword" or "shortened sword" (because placing the forward hand on the blade necessarily shortens the fencer's reach). Italian traditions tend to use the phrase "in armed hand."

We should clarify here that talk of shoving a sword into mail is not hyperbolic. Pseudo-von Danzig describes the technique:

> *Note*: the fourth guard for dueling should be done thus: hold your sword by the handle with your right hand and with your left grasp the middle of the blade. And hold it under your right armpit, and set the one quillon in front right on your chest, and hold your point against your opponent.
>
> **Note a Good Lesson**
> Know that you should come into the Fourth Guard from all the other guards as you plant upon your opponent ... if you then truly hit so that you have your point into his harness, then wind your hilt quickly to your chest into the guard, and press thus forward at him, and do not let him escape from your point.
>
> (Von Danzig, 2010, p. 155)

The advice here is essentially to couch the sword under the armpit like a lance once its point has been planted in an opponent's mail links. This permits a fencer to put the whole weight of the body behind a strongly braced shove.

Pseudo-von Danzig's advice to "not let him escape from your point" is a salient reminder that merely planting the point in mail is not likely to result in significant injury. Because of this, an entire subset of armored fencing deals with ways to deter the opponent from couching the weapon and shoving in just this way. Here is a simple example, also from Pseudo-von Danzig:

> Or thrust outside and behind into the gauntlet of the arm that holds the middle of the sword. And when the thrust sticks, then run with the hand before you to the lists. Thus you again win the upper hand over his side and gain a great advantage.
> (Von Danzig, 2010, p. 157)

In this technique, Pseudo-von Danzig takes advantage of the hourglass gauntlets of his day, which flared out at the wrist to permit greater range of motion. The play begins with the attacking fencer having already planted the sword in the defender's mail and couched it under the armpit, left hand on the blade. An opponent wearing hourglass and grasping the couched sword from below with the left hand must point the funnel mouth of the left gauntlet toward the opponent's right. Thus, a cool-headed defender can thrust the sword into the attacker's left gauntlet. This will certainly be enough to stop the opponent from pressing forward, and, as the gloss says, even allow the defender to press the attacker back. Other such techniques involve thrusting in or near the fingers to pry the opponent's fingers from the couched sword or simply thrusting into the palm of the forward hand, should the opponent be careless enough to expose it in the rush to shove forward.

But let us suppose that the opponent does manage to complete his thrust. What does this translate to in terms of injury? A very pointy sword might be able to get as much as a half-inch of steel through a link even without bursting it. Depending on how much padding is behind the mail, that might translate to a painful stab or little more than a prick. A good strong thrust should be able to burst a link or two and get 2" to 3" through the mail. That's a pretty serious stab wound, but not necessarily enough to be fatal (how much penetration is necessary to get a vital organ depends on the angle of the stab wound). It's also almost certainly not enough to be instantly incapacitating. Thus, even against a strong thrust by a pointy sword, mail will limit the damage done by the attack and give the wearer time to fight back or retreat to safety.

Of course, an opponent who has suffered a serious stab wound will still be weakened and eventually defeated by blood loss. This is something that the bludgeon-through-the-mail approach doesn't do, which presumably is one reason why thrusting techniques tend to predominate over smashing and bludgeoning ones in period fencing manuals.

There's another obvious reason to prefer the thrust to the bludgeoning cut, of course, when it comes to targeting mail between armor plates. Unless a fencer has deliberately left off certain plates (for instance, Fiore illustrates some of his fencers without spaulders, presumably to improve the range of motion in their shoulders), the gaps in plate armor between plates are usually quite small.

Trying to cut into such a small target—say, the seam of the shoulder between breastplate and spaulder—is difficult to the point of impossibility. Stabbing into those gaps is easier, particularly if one hand is on the blade to help steer the point (which is another reason swords are usually held "shortened" when facing armored opponents).

Let us turn now to a more thorough discussion of armor plates themselves. We should begin by stating that it is effectively impossible to cut through armor plate, no matter how sharp or heavy the weapon. I should hasten to add that cutting through armor plate *can* be done under ideal circumstances. For instance, people sometimes use swords to cut mild steel helmets that are braced against a stand, a tree stump, or the like, as a test of skill. Cutting through armor plate that is not braced, but actually being worn by a fighting individual is not something to count on. The same goes for thrusting. A stout point can certainly be forced through steel under experimental conditions. Forcing it through armor plate in fighting conditions is functionally impossible.

This is not to say, and I do not mean to imply, that striking armor plate with a sword blade is utterly useless. It is not likely to cause serious injury, but strong blows with a sword can still inflict pain (particularly to the extremities), disorient (if struck to the helmet), or beat aside an opponent's weapon. Though little discussed by treatises, cuts with a sword blade against a fully armored opponent can appreciably increase a fencer's chances of success later in the fight. The important point to keep in mind is that, against plate armor, this kind of attack is merely preparatory to a technique that will inflict actual injury.

There's an important exception to be made here for "heavy" polearms such as we discussed in Chapter 11. A stout blow from a poleaxe could indeed defeat even a man in plate armor. We have already seen this play from armizare:

> With a half-turn of this poleaxe I'll take it from your hands ... strike you in the head with it ... I don't believe you will survive this ... you'll likely drop dead to the ground after being struck in the head like this.
>
> (Fiore, 2017, p. 37r)

Fiore also claims that, at least in some circumstances, a poleaxe thrust could penetrate a breastplate:

> I'm the Short Serpent Guard ... This guard delivers a powerful thrust that can penetrate cuirasses and breastplates. Fight with me if you want me to prove it.
>
> (Fiore, 2017, p. 35v)

This is not to say, of course, that every poleaxe attack can bypass armor like tissue paper. It is to illustrate, however, that even a man in very heavy armor needed to afford the blows of such weapons a respect that lesser weapons did not command. Modern armored sparring has shown that even rubber-headed poleaxes can cause concussions through helmets and break fingers through lighter gauntlets (modern armored fencing with poleaxes is usually done with mitten-style gauntlets, which are less dexterous than five-finger gauntlets but significantly more protective).

Less effective than poleaxe strikes, but better known, are pommel strikes from a sword. These are a really minor family of techniques, but I want to spend some time on them, because they occupy a disproportionate place in the public imagination. Some people even seem to think that armored combat was simply an orgy of knights bludgeoning each other with their pommels.

First, let's talk about pommel throwing, popularized by the colorful (but inaccurate) translation, "End him rightly." The Gladiatoria Group includes a play for an armored duel in which the attacking fencer unscrews the pommel of the sword and hurls it at the opponent. This creates a window of opportunity for the attacker to close with sword or spear.

There are several things to note here. The first and most important is that the text doesn't claim that pommel throw itself "ends" anything. This is a technique to gain a moment's distraction in a duel; the throwing fencer must still engage the opponent with sword or spear.

When students encounter this technique, they often wonder how much injury the throw itself would do. This depends, of course, on where the pommel hits the opponent and how good the attacker is at throwing. A sword pommel weighs about a pound (give or take a few ounces), which could injure a hand if thrown vigorously enough, even through a gauntlet. On the other hand, the armor over more vital areas (like torso) can take a bodkin-pointed arrow from a war bow at essentially point-blank range. Thus, the risk of injury is present (which is probably why the technique's listed counter is to deflect the throw with the combatant's dueling shield rather than just let it hit), but not likely. Eating a one-pound fastball is likely to be very uncomfortable and make a thunderous great noise, so it is easy to see how it would provide a moment of distraction in which the throwing fencer can attack with advantage. Overall, however, we should not imagine this to be much more than a bit of armored esoterica.

There is one last point to make about pommel throwing: medieval sword pommels do not ordinarily unscrew, and even later-period sword pommels that do unscrew are generally secured by a nut that cannot be unscrewed by hand. Thus, this technique requires buying a specially-made trick sword to begin with, which (judging by the dearth of surviving antiques) virtually nobody actually did.

Let us then turn from pommel throwing to the other way of hitting people with pommels that gets too much press: holding the sword by the blade and using it like a sledgehammer. This technique goes by many historical names; *mordhau* and *mortschlag* (both meaning, roughly, "murder strike") are probably the best well known.

Names like "murder strike" imply that striking with the pommel, sledgehammer-fashion, is an absolutely devastating technique. And it certainly isn't useless. However, some of its other, less well known, names (like "percussive point") give a more balanced idea of its effectiveness.

The basic problem with this technique is that swords are not upside down hammers. A few key differences:

First, when gripping a sword blade, a fencer cannot let the weapon slide through the hands during the strike as is done with a staff or other polearm. Letting the blade slide through the hands in this fashion strike would slice the hands open (even with Monte's left hand strips of mail, parts of the left palm, the undersides of the left fingers, and the entire right hand are still vulnerable to being cut).

Second, a poleaxe is typically at least as tall as a man's shoulder and weighs in the neighborhood of 5 to 7 pounds. A longsword is both shorter and lighter (typically not even reaching the armpit, and weighing 3 to 3.5 pounds); other swords are shorter and lighter still. Thus, the absolute length of the lever arm created by gripping the sword blade is still shorter than that of a true two-handed bludgeon.

Third, sword blades are, of course, made of metal; polearm shafts are made of wood. Consequently, less of the sword's mass is concentrated at the hilt end of the weapon.

Fourth, while pommels can be (and frequently were) designed to concentrate the force of a pommel strike on a small area, it's very difficult to give them the sort of "clawed" or "meat tenderizer" faces that actual warhammers or poleaxes used to avoid slipping off of armor. Remember that part of what makes armor plate good is its shape, designed to cause blows to glance off and thus impart less force to the wearer.

Real hammers have ways to get around this through the shape of the hammer face; pommels don't.

All of this means that an upside-down sword hits a lot less hard than does an actual warhammer. That's not to say that the "percussive point" is useless, but it is important to understand what it does and doesn't do. As with the pommel throw, a pommel strike to fragile extremities or joints can absolutely cause serious injury. No fencer wants to catch a mortschlag to the hand, or the elbow, or the knee. However—as with the pommel throw—the same strike to the head or torso is mostly just going to be stunning and disorienting. That's far from useless, but it needs to be capitalized upon. It isn't a complete attack in itself.

It's also worth pointing out that the percussive point can go catastrophically wrong. Pseudo-von Danzig again:

> *Note*: when you have your sword over your left knee in the guard and he isn't strong, and strikes with the pommel to your head, then catch the stroke in the middle of the blade. And sending your pommel outside over his sword near behind the hilt, pull with it downward to your right side. Thus you will take his sword; and also plant upon him.
>
> (Von Danzig, 2010, p. 158)

In this technique, the defending fencer parries the incoming pommel strike on the sword, between the fencer's hands. The defender's pommel snakes over the parried blade, to the left. The defender then pulls strongly down to the right, under the opponent's blade. This has the effect of hooking the mortschlaging sword and pulling it out of the opponent's hands. Grisly damage to the opponent's hands ensues unless the opponent drops the sword with alacrity.

In other words, while there are options for using a weapon directly against armor plate, none of them are ideal. All of this is why fencers apparently worked so hard to work around an opponent's plates, rather than simply mortschlaging each other all day long.

This brings us to our last and most important family of anti-armor techniques: grappling. We have already seen examples of using a sword as a lever to help force an opponent to the ground, or as a hook to imbalance and trip.

The importance of throwing an armored opponent to the ground cannot be overstated. It is extremely difficult to reach the vitals if an armored opponent is still standing. Plate armor cannot keep a fencer completely safe from harm (recall the glosses we have seen about finding an armored opponent's openings), but it does an excellent job of defending the vitals. Thus, while it is always possible to kill an armored opponent with a good poleaxe blow to the head or a devastating thrust to the face through an open visor, for the most part, plate armor prevents such one-shot kills. Actually killing an opponent in plate armor usually required pinning the opponent simply in order to reach the required parts of the anatomy.

The percentage of techniques that end in wrestling is vastly higher in armored fencing than in unarmored fencing. It is the very nature of armor to make weapons less effective than they otherwise would be, after all, and the less effective weapons become, the more the fight resembles a contest between unarmed combatants. It is no wonder that

238   THE USE OF MEDIEVAL WEAPONRY

Liechtenauer advises any "young knight" to "wrestle well." Indeed, Pseudo-von Danzig reminds his readers not even to bother drawing their daggers before winning the grappling contest:

> Now you should know that, for the most part, all fighting in single combat in harness comes in the end to dagger fighting and to wrestling. Therefore note, when you close with an opponent, then attend to nothing else but the wrestling and let your dagger stay in its scabbard, because you cannot hurt him through the harness as long as he is standing before you and hinders your hand. When you secure him with the wrestling or have thrown him and have overcome him, then work with the dagger to the openings.
>
> (Von Danzig, 2010, p. 158)

This brings us back to an oft-overlooked point about the importance of thrusting: it permits a fencer to control an opponent's movements in a way that percussive blows do not. As we discussed earlier in this chapter, a thrust into mail is not likely to penetrate deeply enough to kill in its own right. However, an armored opponent cannot step forward against a planted point—nobody wants to push their whole armored bulk *against* a point that is already sticking into them! In the majority of cases, a thrust is only the beginning of an attack, even if it connects and makes the opponent bleed. The real point of the thrust is to constrain the opponent's momentum so that a skilled grappler can exploit that momentum. Pseudo-von Danzig again:

> If he wants to step back and flee from the point or spring backwards away from it, or turn out from the thrust, and turns his side toward you, then spring to him and see that you grasp him safely and truly with an arm break or else with other wrestling techniques.
>
> (Von Danzig, 2010, p. 146)

The importance of putting an armored opponent on the ground is true for all weapons, even poleaxes and other such heavy polearms. My favorite illustration of this principle comes from the Anonimo Bolognese. As we alluded to briefly in Chapter 11, the Anonimo Bolognese includes a number of techniques that end with a poleaxe thrust to the groin of an armored man. Here is one such. In this play, the attacking fencer has set up the technique by harassing the opponent's feet with thrusts from the butt-spike, causing the opponent to likewise parry with the butt of the axe. The play continues:

> [S]trike the head of your poleaxe into the heel of his with the greatest possible force, so strongly that he will, without fail, remove his left hand from his poleaxe. Then pass forward with your right foot and renew your attack by striking to the part of his groin above his testicles with your spike, using all possible force. Then raise your poleaxe, shove him backwards, and throw him forcefully to the ground.
>
> (Anonimo, 2020, p. 237)

The attacking fencer delivers a heavy blow from the head of the axe, knocking the opponent's weapon halfway out of the hand. The attacker then gives a rising thrust with the top spike of the axe to the opponent's genitals "with all force." This is an undeniably brutal technique … but even this poleaxe-uppercut to the groin is ultimately just another way to throw the opponent to the ground.

Both of these techniques continue a theme in this chapter: thrusting is the most powerful and versatile technique against armor. In an armored contest, thrusting offers the greatest wounding potential, works against the widest variety of targets, and best sets up an opponent for a fall. Smashing an opponent in armor with a bladed weapon, whether with the pommel or with strong blows from the sword, comes in a very distant second. These attacks have their uses, but they are much less likely to lead the fight toward its ultimate conclusion on the ground, where a real fatal injury can be delivered.

This raises an important point about armor and injury. Modern people often subconsciously think of medieval armor in terms of its offensive capabilities. We imagine armored knights smashing their way through enemy lines, and ask questions like how many "peasants" a "knight" can take on at once. Thinking of armor in this way overlooks a major defensive benefit: it makes the wearer more survivable.

One of the ways armor does this is by limiting the damage a fencer can receive while wearing it. Getting a 3" stab wound from a sword or spear is very unhealthy, but it is far better than being run all the way through. Likewise, as we have discussed, even the gaps in armor generally make it hard for a point to find a vital organ. Lastly, while being smashed by a pommel, sword blade, or axe may be disorienting, painful, and even inflict a concussion, those are all far preferable to the consequences of being hit by a blade or pommel *without* wearing armor.

Survivability literally means a fencer is more likely to survive. A fencer can recover from a concussion. It is much harder to recover from a cut that splits the skull. A stab wound is bad. It is not as bad as having a limb lopped off.

We often think of armor in purely military terms, but for the people who wore it, it was intensely personal. Armor did allow a fencer to press on through blows that would have felled a less well-equipped man, to be sure, with obvious positive effects for the overall course of a battle. But it also often meant the difference between being functional enough to retreat after being wounded or being butchered on the spot. Survival, not invulnerability, was armor's first goal.

But suppose that an armored opponent could not (or would not) withdraw? Absent retreat or surrender, how could an armored man be killed? There are, essentially, two scenarios.

The first scenario is that a fencer manages to land a truly fatal attack despite the opponent's armor. Lucky blows to the head with a poleaxe and stabs to the face (whether after wrenching open a visor or against a fencer whose visor is up or absent) fall into this category. A thrust to a gap might also penetrate deeply enough to kill under unusual circumstances, such as if there is no mail covering that gap or if the mail suffers from some hidden flaw that causes it to fail catastrophically.

The second scenario is that the fencer does not land a lucky fatal blow. In that scenario, the fencer must bring the opponent to the ground, pin the opponent on the ground, and remove enough armor to stab the opponent to death with a dagger.

Again we return to survivability. It is, of course, extraordinarily difficult to pin a struggling man in armor and remove enough pieces of armor to get access to a vital area.

This may surprise you. Some modern people imagine that armored fencers were helpless on the ground, or even that they could not easily rise due to the weight of their armor. On the contrary, ground fighting in harness is not only possible but an important subset of a fighting man's skillset. An armored opponent on the ground is not an overturned steel turtle; far from it. Even a thrown opponent must be treated as a dangerous grappler, and probably still has a dagger, to boot (still in its sheath, if the opponent followed Pseudo-von Danzig's advice).

This raises an uncomfortable prospect. What is a fencer to do in an armored contest if it turns out that the opponent is simply the superior grappler? Why bother to use weapons at all in an armored fight, if the contest "for the most part" ends in grappling?

The answer is that weapons—even if they do not inflict fatal wounds—allow a fencer to change the odds of the final desperate grapple. Even the largest, strongest, and most skillful grappler can be out-wrestled if suffering from multiple stab wounds and blood loss (perhaps with a concussion or smashed fingers thrown in for good measure). Every injury that inflicted upon the opponent can prove to be a critical advantage in the wrestling match and ground fight that actually leads to a kill.

Huntzfelt authored a treatise that gives us our clearest look at what an armored ground fight to the death actually looked like. The following excerpts speak for themselves:

> If you throw him onto his stomach:
>> Then fall with your right knee behind in his abdomen, and grasp with your left hand over his head in front on the visor, and pull upwards and press down with your knee.
>>
>> If his visor is open, then grasp with your fingers to his eyes, or grasp with both hands on his helmet and twist his neck. Thus he falls again on his stomach.
>>
>> ...
>>
>> *Item*: When he lies on his stomach, then sit upon him and stride over his arms and break them.
>>
>> ...
>>
>> If you have thrown him under you, then cut or tear a large piece from his arming coat, and thrust it in his visor with the dagger and do not let him free himself from it. This is good when it is dirty. Or throw dust or muck into his visor, or dig out some earth with your dagger and throw it in his face.
>>
>> ...
>>
>> *Note*: If you can find no openings on him, then cut him out of the arming coat at the arms or the hose, and whatever laces you can get your hands on, cut them off. And if you find a belt, then cut that also, whenever you will find that will be helpful to you; and meanwhile look to his dagger.
>>
>> ...

> Thus seek the openings behind to his body, into his face, between the shoulders, inside the armpits, and do not pull back until he surrenders.
>
> (Von Danzig, 2010, p. 177–179)

This is really gruesome stuff, which leads us to our last point on the subject of armored fencing. Not all armored combats ended in death, of course: an armored opponent might retreat once wounded, or surrender. As we have just seen, Huntzfelt contemplates surrender. He does not, however, indicate why a fencer might surrender in the first place. In some parts of Europe, a fencer might hope that the prospect of ransom would excite a victorious opponent's mercy. The nature of armored combat itself, however, suggests another possibility.

Armored combat is intensely personal in a way that even an unarmored fight with hand weapons is not. Swords and polearms can inflict such catastrophic damage against unarmored bodies that the fatal moment may (though it does not always) come swiftly and unexpectedly. Armored fencing is different. Most of the time, in an armored fight, both combatants see the moment of death coming. The stabs and smashes are painful; the blood loss is palpable. On the ground, there is the mad scramble of wrenching at helmets or gouging at eyes or cutting at the straps of armor or trying to stab daggers into testicles. There is, in other words, time to think about what's coming, and how it won't be fast. There is also time to think that, compared to such a death, surrender truly is preferable. There is time for the *opponent* to think about the prospect of stabbing a struggling man to death in the mud, and to consider asking for surrender.

And in giving both fencers time to think about these gruesome realities, their armor might perform one last act to keep them alive.

CHAPTER THIRTEEN

Honor in wars: military fencing

> *Young knight learn*
> *to love God and revere women;*
> *thus your honor will grow.*
> *Practice knighthood and learn*
> *the Art that dignifies you,*
> *and brings you honor in wars.*
> —Johannes Liechtenauer (Von Danzig, 2010, p. 96)

When I first discovered the existence of HEMA in high school, I decided that I was mostly interested in unarmored fencing with sidearms, such as swords. I knew enough medieval history to know that plate armor, polearms, and shields were mostly restricted to the battlefield, and I had no interest—so I thought—in learning techniques for a type of warfare so far removed from anything I was likely to experience in my daily life.

Not everybody who comes to HEMA feels the same way. Plenty of historical fencers are keenly interested in learning techniques that would be appropriate for the medieval battlefield, and it is not uncommon for these fencers to ask whether our surviving fencing treatises really have anything at all to do with military fencing.

Perhaps you have the same question. Even at public-facing demos, I occasionally encounter an audience member who knows a bit about HEMA and has somehow come by the notion that our surviving treatises are really nothing more than dueling manuals, thoroughly removed from the realm of warfare.

This is untrue, and we should address the misconception now. A full treatment of medieval warfare is beyond the scope of this book, and this chapter will not attempt to give one, but it is important that we understand how our surviving treatises fit into the world of war.

There are three issues we need to address in this chapter. The first is that "military-grade fencing" is not a medieval concept. The second is the misconception that our surviving treatises do not explicitly concern themselves with war. The third is the window of opportunity, even in organized combat, during which a combatant must kill or be killed.

To begin with the first: in the medieval milieu, there is no such thing as "military" fencing as distinct from "civilian" fencing. This dichotomy is an anachronism imported from modern preconceptions. Some people seem to think that when a medieval person joined the Army, a grizzled sergeant (or perhaps their equally grizzled veteran comrades) taught them how to handle a spear or sword the Army way.

The problem with this notion is that in medieval warfare, there *was* no Army. There were certainly *armies*, but the modern Army with a capital A—an organized military machine with persistent institutional knowledge—simply did not exist. Still less existed the notion that it was the Army's job to train its members in the use of their weapons. Early modern military manuals, when they exist, concern themselves not with how to train men in the regulation version of cut and thrust but with how to maneuver large bodies of men without becoming hopelessly tangled. Indeed, true "military" HEMA manuals—that is, treatises whose topic *is* how to teach men to fence "the Army way"—don't enter the historical record until the late 18th century. Throughout the Middle Ages, and even throughout the Renaissance, and for centuries even after that, the only way a man had to learn the nuts and bolts of how to use his weapons was through private training.

This being the case, we might expect to see at least some of our fencing treatises concerned with the military use case. And indeed we do. Let us examine a few of the ways how.

It is a common misconception, even among HEMA practitioners, that our treatises do not explicitly discuss the use of weapons at war. This is flatly untrue. Consider the opening of Liechtenauer's zettel:

> *Young knight learn*
> *to love God and revere women;*
> *thus your honor will grow.*
> *Practice knighthood and learn*
> *the Art that dignifies you,*
> *and brings you honor in wars.*
>
> (Von Danzig, 2010, p. 96)

Here, Liechtenauer explicitly situates his teachings with respect to war. That is clearly not his only concern—even in these few lines, he is also concerned with the general concept of "knighthood" and the specific chivalric concepts of honor and courtly love—but it is an explicit concern.

Of course, we might suspect that these lines of the zettel are mere puffery—fancy advertising copy by an itinerant swordsman who needed to sound tough in order to attract students. Are we really just supposed to take his word for it that his art is useful in war?

Yet Liechtenauer's words fit into a larger pattern in the medieval treatises. Writing at approximately the same time as Liechtenauer, Fiore writes in his introduction:

> Even so, my desire for this exercise declining, and so that so much military experience (which furnishes a most valid sustenance to expert men in warfare or in any other tumult) be not lost negligently, I have decided to compose a book regarding the most useful elements of this splendid art, putting in it various figures with examples, by which methods of attack and defense and parries can an astute person be served in fencing or pugilism.
>
> (Fiore Pisani-Dossi, 2019)

Fiore is quite explicit that his book was written in order to preserve his and his teachers' *military* experience—and yet, the text itself concerns nothing but the nuts and bolts of how to use medieval weaponry in personal combat. The only reasonable conclusion is that Fiore considered his art to *be*—among other things—military fencing.

In a similar manner, Manciolino says in the final line of his text that his book comprises "chapters or general rules on the gallant and warlike Art of Fencing" (Manciolino, 2010, p. 145). Marozzo, too, opens his treatise by claiming that it is about "situations arising in the military art" (Marozzo, 2018, p. 73). He elaborates, in the introduction to his first book:

> Although military discipline and art may be noted plainly and clearly in many courageous knights and greathearted fighters, it can also be seen that they are unclear to many very talented people, owing to their inexperience ... I was compassionately moved to exert my wits and my art in order to advise these bold fighters so that they might take up arms properly, for I have seen some vigorous and heroic men be overcome by those less powerful than them, which came to pass for no other reason than through an error on their part.
>
> (Marozzo, 2018, p. 77)

That is to say, one of the reasons Marozzo is writing is to help talented but inexperienced people acquire "military discipline and art." In the next paragraph he clarifies that he is not speaking of single combat only:

> Accordingly, above all else, I exhort, indeed I admonish, anyone who may be about to enter into single combat, *or into battle*, to have justness on his side (emphasis added).
>
> (Marozzo, 2018, p. 77)

Godinho, too, makes reference to the military application of his art. As we saw in Chapter 8, for instance, he includes a rule for using the greatsword aboard a warship when boarders have taken control of both the stern and the bow of the ship. In a theoretical discussion of how agility is not the same thing as speed, he also makes mention of direct military experience by a trained fencer, or *diestro*:

The *diestro* notes that this setting off doesn't always have to be with agility, as many of us have seen in battles, as well as in individual fights, because they throw themselves in without consideration, they fall in the case of ambushes, for which occasion those on horse have the captain, and those on foot have the sergeant.

(Godinho, 2016, p. 121)

Lastly, dall'Agocchie makes this impassioned defense of the military applicability of fencing:

> GIO: In general, then, (as I told you) one takes it to refer to any sort of military practice, since the military art consists of nothing other than knowing how to judiciously and prudently defend oneself from the enemy, and harm him, whether in cities, or in camps, or in any other place. Since this word, "fencing," means only to defend oneself with a means of harming the enemy, clearly it can be taken generally for any kind of combat.
>
> But taking it specifically, for one-on-one combat, it is manifestly clear that it is part of, or rather a ladder and a guide to, the art of war, as many times it is necessary to employ this art in defense of one's own life, as in examples that one reads in so many histories, or that one sees every day. Therefore I tell you that one cannot be said to have a basis in, nor be perfect in, the military art if he lacks this portion of it, given that nothing is called perfect if it is lacking, or it can be augmented; and if one has to add to the art of combat the knowledge of how to defend his own person, which is indeed its fundamental principle, then, lacking this art, he cannot ever be called "perfect."

(dall'Agocchie, 2018, p. 4)

Nevertheless, if the only connection our treatises had to "military" fencing was bare insistence that these works pertained to the use of weapons on the battlefield, we might yet cherish some skepticism. Perhaps this is *all* mere advertising puffery. Perhaps, we might think, claims that an author's lessons have military utility is a mere formality required by the genre. Perhaps, we might think, all these authors were honestly deluded!

I have a difficult time crediting such skepticism. For one thing, it is difficult to imagine a civilian context in which some of the situations in our treatises might occur. As I said in Chapter 1, I have deliberately omitted any discussion of mounted combat to this point, but we must briefly adduce some mounted examples here. Consider this curious paragraph from Fiore:

> This scoundrel was fleeing from me towards a castle. I rode so hard and fast at full rein that I caught up with him near his castle. Then I struck him with my sword in his armpit, which is a difficult area to protect with armor. Now I'll withdraw to avoid retaliation from his friends.
>
> (Fiore, 2017, p. 46v)

Fiore appears to be recounting an actual personal encounter here. Is this a dueling scenario, or even one that might arise during a civilian dispute (why is the fleeing victim wearing armor?) Likely not. Talhoffer's 1467 manuscript, produced for the graf von Württemberg,

HONOR IN WARS 249

includes four illustrations concerning the use of a crossbow by a man in partial armor from horseback against one or multiple lance-armed opponents. This, too, is a difficult inclusion to square with a purely civilian or dueling context.

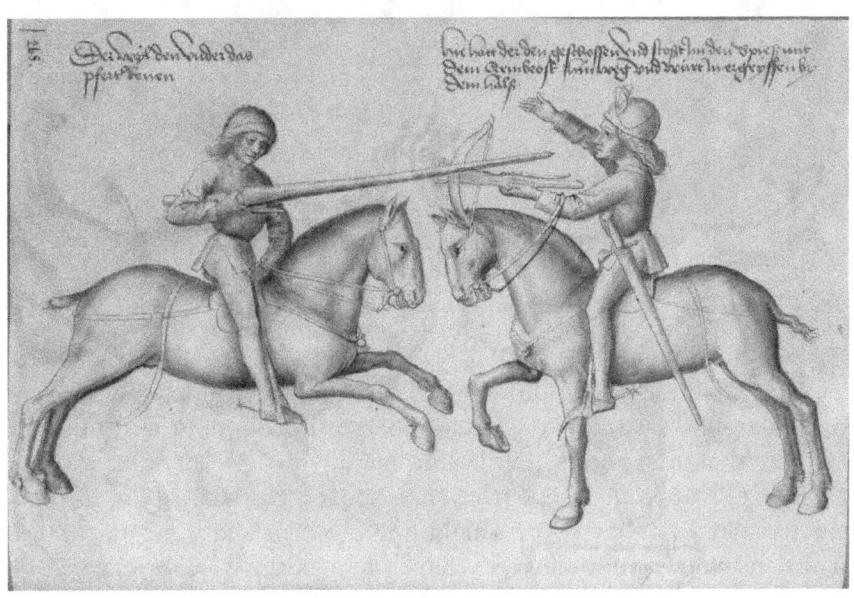

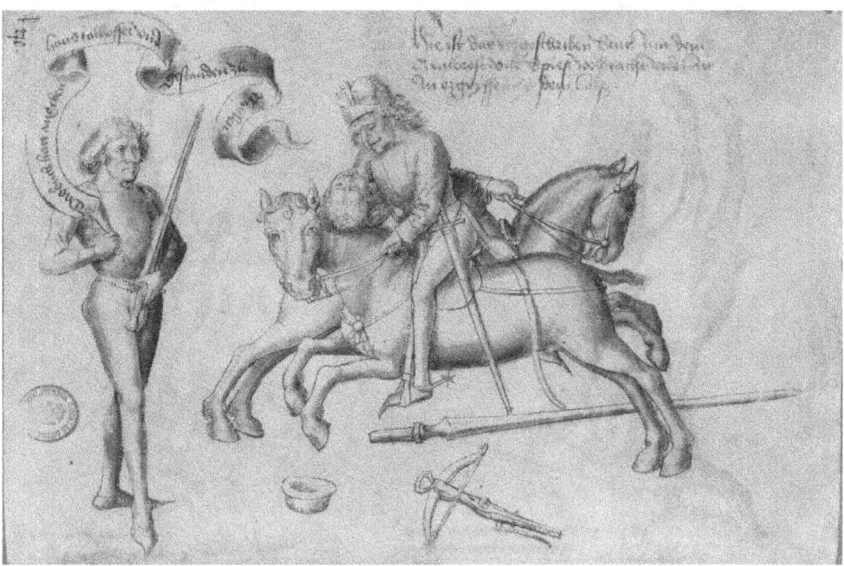

The Bolognese are no different. Marozzo's two-handed sword techniques are, he says, for use "one-on-one or in company, and in every fashion that may occur while employing the two-handed sword" (Marozzo, 2018, p. 206). His sword and spear material includes the possibility that "you have your shield on your arm and your partisan in hand and you have to defend yourself in a skirmish" (Marozzo, 2018, p. 270). He includes, moreover, separate discussions of how to fence with a sword and shield against polearms, a sword in two hands against polearms, and different kinds of polearms against each other. Asymmetrical encounters like this do not fit the dueling context, and they are too heavily armed to easily imagine arising other than on the battlefield. For that matter, the mere fact that both Marozzo and Manciolino discuss the use of the pike—the late medieval and Renaissance formation weapon *par excellence*—ought to arouse suspicion that they are not confining themselves to the dueling ground or the street fight.

We may, therefore, dismiss the notion that medieval martial arts had nothing to say about wartime fencing. Two possibilities remain. The first is that they contain only smatterings of military-applicable techniques, that *only* the examples I have adduced in this chapter, and others similarly incontrovertibly military, should be credited as having currency on the battlefield. And certainly, it must be admitted that the *incontrovertibly* military references are scattered throughout our extant treatises somewhat haphazardly.

The second possibility, though, is to take our authors at their word: that they really did mean what they said in their introductions, and that their entire traditions were military-applicable. In this reading, the unambiguously "military" techniques only *appear* to be scattered haphazardly throughout the treatises, and only if we insist on disbelieving our authors when they insist that the art of fencing genuinely is a *military* art. I find this second reading much more persuasive.

I grant, though, that even on this second reading, it can be difficult to imagine how some of the techniques in our fencing treatises would translate to the battlefield. Consider the discussion of armored combat from Chapter 12, for instance. Was it really possible to go through the complicated process of wounding an armored man sufficiently to bear him to the ground with a multi-step grappling technique and then pin him, find an opening in his armor, and stab him to death with one's dagger ... all in the midst of battle? Can a man really use wide, sweeping techniques such as the zwerchau while in formation without fouling the blades of his allies (or worse, striking his allies themselves)? Did anybody even have the mental wherewithal to employ complicated martial arts techniques in the brutal violence of medieval battle ... and if they did, would such subtleties be effective?

Dall'Agocchie discusses this latter objection directly. Standing in for the reader, as usual, Lepido Ranieri asks:

> LEPIDO: [C]ertain doubts remain which I'd like you to clarify for me (before we move on), and one of them is this: there are many

who say that when acting in earnest one can't perform so many subtleties as there are in this art.

GIOVANNI: What do they mean by "subtleties"?

LEP: They say that one can't feint, nor disengage, and that there isn't enough time to perform body evasions and similar things.

GIO: They say such things because one rarely finds men who aren't moved by wrath or fear or something else when it comes to acting in earnest, which causes their intellect to become clouded and for this reason they can't employ them. But I say to you that if they don't allow themselves to be defeated by these circumstances, and they keep their heads, although they may be difficult, they'll do them safely.

LEP: But what's the reason for teaching them if they're so difficult to employ in earnest?

GIO: They're taught so that courageous men can avail themselves of them in the appropriate occasions. Because one often sees many who were somewhat timid and fearful, yet nonetheless were able to perform them excellently when done in play; but then they were unable to avail themselves of them when the occasion arose in which to do them in earnest.

(dall'Agocchie, 2018, p. 40)

Dall'Agocchie here makes a point that is important to keep in mind about medieval martial arts. These arts were not intended to permit the weak and timid to prevail, through sheer skill, over the strong and bold. As dall'Agocchie says, the skills of the martial artist are "taught so that *courageous* men can avail themselves of them in the appropriate occasions" (emphasis added). Indeed, dall'Agocchie says, a timid but skilled fencer will quickly find himself emotionally overwhelmed in actual combat and unable to perform any of the techniques that he theoretically "knows."

While modern sparring experience is not a direct analog to earnest combat with live weapons (nor can it be), it is close enough to confirm what dall'Agocchie says, and to add significant weight to the words of other authors who exhort their students and readers to be bold, courageous, or carefree when fencing. Many HEMA students have had the experience of transitioning to sparring for the first time after months of drill, only to discover that even the chaos of *mock* fighting drives every single technique they know straight out of their heads. It is only when a student can face an opponent without fear or over-excitement that the techniques and principles of fencing can truly be applied—even in sparring. How much more so in earnest combat?

That said, modern experience also suggests that *when* a fencer has mastered his emotions during a bout, historical techniques are indeed effective. One of the most valuable experiences a modern historical fencer can have is to take historical techniques outside the HEMA world to spar with people who do not necessarily study historical European martial arts, such as lightsaber fencers or live-action role-players. While the rules of those communities may force certain compromises (many LARP groups disallow attacks to the head, for instance, which is a significant handicap for historical techniques), it is invaluable to be able to practice the "subtleties" of medieval martial arts against opponents who are often skilled fencers in their context but have not studied historical European sources. In my own experience and that of my students, the success rate of most medieval techniques jumps markedly against opponents who are not immersed in the HEMA milieu. This suggests that skilled medieval fencers who had the presence of mind to remember their training in the press of battle would indeed have found their art effective against most opponents.

All this said, it must be admitted that not all techniques are equally applicable to all military scenarios. A sweeping horizontal longsword cut like a zwerchau, for instance, requires significant clearance to the fencer's left and right and is not especially effective against an armored opponent. Likewise, many of the relatively weak cuts to the hand that can be performed in sword and buckler fencing are fairly ineffective against armor.

There are two things to say about this. The first is that, while not all *individual* techniques translate well to military fencing, the fundamentals of all the martial arts systems we have touched upon in this book *do* translate. Second, and perhaps more often overlooked, is that "military" medieval combat is not necessarily comprised solely of set-piece battles in which large formations of men engage other large formations of men in open terrain. Indeed, while such encounters may capture the imagination, they were quite uncommon in the annals of medieval warfare. A medieval soldier would much more likely be called upon to employ his weapons when defending or storming a fortification, while raiding enemy territory, or in other small skirmishes, than in a massive battle surrounded by the press of hundreds or thousands of men.

Even in such set-piece battles, formal fencing training would be applicable. This may be unintuitive, as the art of fighting in formation is first and foremost about march discipline. As anybody who has participated in a melee exercise of any scale can readily attest, a disciplined and coordinated line of fencers is far more effective than an equal-sized group of fencers acting as individuals, even if the undisciplined side has greater skill at arms. This fact is only exacerbated as the number of combatants increases. When the number of participants is large, the side with a better ability to move troops on the battlefield without devolving into a chaotic mob has a commanding advantage.

And yet, the moment comes even in large combats when a fencer must employ his or her weapon to strike an opposing combatant, or use his or her arms to avoid being struck. All that marching is meaningless if the troops can't actually use their weapons, after all. It is in these moments when fencing training comes to the fore. Recall the halberd vs. greatsword matchup we discussed in Chapter 11, and Marozzo's discussion of the skillful and unskillful polearm fencer. His assumption was not that the skillful polearm fencer would employ numerous fancy techniques or great space-hogging wheeling cuts, but rather that the fencer's skill would translate into harder, faster strikes that were less likely to overcommit. These are benefits of fencing training that are of material value even in a close-order formation and even in very large battles.

It is easy, when discussing historical martial arts or reading the treatises themselves, to become fixated on technique. And no wonder: there is only so much an author can say about the fundamentals of timing, distance, and striking form. But we should not be seduced by this fact into the error of thinking that historical fencing (or any fencing, for that matter) is mostly about building a robust library of techniques. It is *mostly* about improving a fighter's fundamentals. For this reason, all other things being equal, during the brief moments in battle when an enemy is within reach of her weapon, a trained fencer is more likely than an untrained fencer to successfully bring down her man. Likewise, the trained fencer is more likely to be able to fend off the attacks of the enemy that are directed at her, and less likely to unconsciously do something in the line of battle that might leave her momentarily exposed.

There are two other benefits to individual training in battle as well. Precisely because a well-trained fencer is more likely to be able to cause casualties among the enemy and better able to avoid becoming a casualty himself, fencing training improves a soldier's confidence and morale. Confident men maneuver about the battlefield with greater alacrity than unconfident men, and are more willing to close within spear-reach of the enemy. Thus, fencing can positively influence march discipline—and, thus, the outcome of the battle—even if a fencer never has the opportunity to strike an enemy soldier personally.

And so, though it may seem surprising at first, we are justified in treating historical fencing treatises as applicable to "military" contexts. If an understanding of those treatises does not square with our preconceptions about what medieval battle looked like, our response should not be incredulous disbelief, but the joy of discovering the unexpected. For all of us who are interested in the martial arts of the past, this is only fitting. If not for the joy of discovering history, none of us would be here.

CHAPTER FOURTEEN

Final thoughts

We come now to the close of our survey of the use of medieval weaponry. Through the works of over a dozen period fencing masters, we have explored the historical use of the longsword, sword and buckler, sword and shield, sword and cloak, sword and dagger, two swords, dagger, greatsword, spear, and poleaxe, halberd, and bill, in and out of armor. We have examined how these weapons were used by fencers to address the problems that confront an individual in armed combat, the idiosyncrasies of each weapon combination, and how different medieval martial arts systems employed them. Along the way we have looked at many plays from historical treatises and—I hope—come to appreciate the technical depth and complexity of medieval fencing.

Martial arts have a way of challenging our preconceptions, and historical European martial arts are no exception. When I began my HEMA journey, I only wanted to learn how to use a sword, and the traditional martial arts of medieval Europe seemed like a practical, no-nonsense way to do that. But the longer I practice, the more I find myself motivated by the traditions themselves: to reconstruct and preserve something beautiful, and to let it be seen again. If this book has dispelled any of your preconceptions about the fencing of the past (perhaps beginning with the word fencing to describe historical fighting!)—if it has helped you see these traditions more clearly—then it has been a success.

In Chapter 1, I admitted that this book exists because swords are cool. But there is another reason as well: the excitement of bringing the past alive, and perhaps making fantasy just a little more real. Let that be the through-line for this book: to begin with the love of swords and medieval weaponry, and, by discussing the works of medieval fencing masters, to make that fantasy just a little more real. Historical fencing masters often included in their treatises a hope or wish for the uses to which their work would be put—to preserve their experience and the knowledge of their teachers, perhaps, or to train future generations of young (fighting) men in true skill and good morals. This is mine.

WORKS CITED

Anonimo Bolognese. *With Malice and Cunning: Anonymous Treatise on Bolognese Swordsmanship*. Translated by Stephen Fratus. Self-published, 2020. (cited as Anonimo)

Dall'Agocchie, Giovanni. *The Art of Defense: on Fencing, the Joust, and Battle Formation*. Translated by W. Jherek Swanger, Self-published, 2018. (cited as dall'Agocchie)

De'i Liberi, Fiore. *The Flower of Battle: MS Ludwig XV 13*. Translated by Colin Hatcher, Tyrant Industries, 2017. (cited as Fiore)

De'i Liberi, Fiore. "De'i Liberi, Fiore." *Wiktenauer*. http://wiktenauer.com/wiki/Fiore_de%27i_Liberi. Translated by Michael Chidester, accessed 10/31/2019. (cited as Fiore Pisani-Dossi)

Godinho, Domingo Luis. *Iberian Swordplay: Domingo Luis Godinho's Art of Fencing*. Translated by Tim Rivera, in collaboration with Steve Hick, Eric Myers, Manuel Valle, and Jaime Girona, Freelance Academy Press, 2016. (cited as Godinho)

Kal, Paulus. "Paulus Kal." *Wiktenauer*. https://wiktenauer.com/wiki/Paulus_Kal. Translated by Christian Henry Tobler, accessed 10/31/2019. (cited as Kal)

Leoni, Tom. *The Complete Renaissance Swordsman: A Guide to the Use of All Manner of Weapons—Antonio Manciolino's* Opera Nova (*1531*), Freelance Academy Press, 2010. (cited as Manciolino)

Marozzo, Achille. *The Duel, or The Flower of Arms for Single Combat, Both Offensive and Defensive*. Translated by W. Jherek Swanger, Self-published, 2018. (cited as Marozzo)

Monte, Pietro. *The Collection of Renaissance Military Arts and Exercises of Pietro Monte: A Translation of the* Exercitiorum Atque Artis Militaris Collectanea. http://mikeprendergast.ie/monte/. Translated by Mike Pendergrast and Ingrid Sperber, accessed 10/31/2019. (cited as Monte)

Nuremberg Hausbuch. Translated by Christian Trosclair. Personal correspondence.

Sigmund ain Ringeck. Translated by Christian Trosclair. Personal correspondence. (cited as Ringeck)

Talhoffer, Hans. "Hans Talhoffer/Württemberg." *Wiktenauer*. https://wiktenauer.com/wiki/Hans_Talhoffer/W%C3%BCrttemberg. Translated by Cory Winslow, accessed 10/31/2019. (cited as Talhoffer)

Tobler, Christian Henry. *In Saint George's Name: An Anthology of Medieval German Fighting Arts*. Freelance Academy Press, 2010. (cited as Von Danzig)

Viggiani, Angelo. *The Fencing Method of Angelo Viggiani: Lo Schermo, Part III*. Translated by W. Jherek Swanger, Self-published, 2018. (cited as Viggiani)

Additional reading

"Gladiatoria Group." *Wiktenauer*. https://wiktenauer.com/wiki/Gladiatoria_group. Translated by Benedict Haefeli, accessed 10/31/2019.

Monte, Pietro. *Collectanea: The Arms, Armour and Fighting Techniques of a Fifteenth-Century Soldier*. Translated by Jeffrey L. Forgeng, The Boydell Press, 2018.

Treatise images

Images from Achille Marozzo, Res/4 Gymn. 26, appear courtesy of the Bayerische Staatsbibliothek on pages 43 (Bayerische Staatsbibliothek München, Res/4 Gymn. 26, folios 26, 36, and 49), 55 (Bayerische Staatsbibliothek München, Res/4 Gymn. 26, folio 83), and 139 (Bayerische Staatsbibliothek München, Res/4 Gymn. 26, folio 308).

Images from Fiore de'i Liberi, MS Ludwig XV 13, appear courtesy of the Getty's Open Content Program on pages 23, 30, 32, 35, 226, and 248.

Images from Fiore de'i Liberi, the Pisani-Dossi manuscript, appear courtesy of the Wiktenauer on pages 27, 135, 136, 137, 140, 173, 191, 231, 233, and 237.

WORKS CITED   261

Images from Hans Talhoffer, Cod.icon. 394 a, appear courtesy of the Bayerische Staatsbibliothek on pages 54 (Bayerische Staatsbibliothek München, Cod.icon. 394 a, folio 117r), 63 (Bayerische Staatsbibliothek München, Cod. icon. 394 a, folio 121v), 83 (Bayerische Staatsbibliothek München, Cod. icon. 394 a, folio 79r), 88 (Bayerische Staatsbibliothek München, Cod.icon. 394 a, folio 54r), 89 (Bayerische Staatsbibliothek München, Cod.icon. 394 a, folio 69r), 90 (Bayerische Staatsbibliothek München, Cod.icon. 394 a, folio 68v), 92 (Bayerische Staatsbibliothek München, Cod.icon. 394 a, folio 57r), 93 (Bayerische Staatsbibliothek München, Cod.icon. 394 a, folio 57v), 207 (Bayerische Staatsbibliothek München, Cod.icon. 394 a, folio 50r), 249 (Bayerische Staatsbibliothek München, Cod.icon. 394 a, folios 135r and 135v), and 250 (Bayerische Staatsbibliothek München, Cod.icon. 394 a, folios 136r and 136v).

Images from Hans Talhoffer, Ms.Thott.290.2°, appear courtesy of The Royal Danish Library on pages 28 and 103 (Royal Danish Library, Thott 290 folios 156 and 249), 57 (Royal Danish Library, Thott 290 folio 243 and 248), 63 (Royal Danish Library, Thott 290 folio 241), 73 (Royal Danish Library, Thott 290 folio 198), and 74 (Royal Danish Library, Thott 290 folio 200).

Image from Paulus Kal, Cgm 1507, appears courtesy of the Bayerische Staatsbibliothek on page 91 (Bayerische Staatsbibliothek München, Cgm 1507, folio 98).

INDEX

Aketon: *see* Armor, Arming Clothes
Anonimo Bolognese:
   Armor, 238
   Cutting Polearms, 205, 207, 208,
      213, 214
   Greatsword, 148, 149, 152
Antonio di Luca, Guido, 10
Arming Sword, definition, 42–43
Armizare, 11, 14
   Armor, 219, 225, 231, 232
   Cutting Polearms, 204, 210, 212, 214
   Dagger, 131, 132, 133, 134, 135, 136,
      138, 140, 144, 145
   Longsword, 17–18, 19, 20, 21, 22,
      26, 27, 29, 31, 32, 33, 34, 37
   Mace, 173, 174
   Spear, 181, 182, 185, 191, 192,
      193, 194
   Warfare, 245, 246, 248
Armor, 39, 67–69, 95, 199–200, 219–242
   Arming Clothes, 221, 223, 224
   Blunt Impact, protection from, 95,
      175, 223, 228, 232, 234–236, 239

Configurations, effects of, 222,
   227, 231
Cuts, protection from, 223, 224–225,
   227, 232, 239, 253
Firearms, protection from, 220
Grappling, 237–238, 240–242
Half-Sword Grip, 229
Incapacitation, 219, 230–231, 234,
   236, 239–240, 242
Leather, use as armor, 221–222
Linen, defenses offered by, 221,
   224–225
Mail, defenses offered by, 221,
   227–231
Mail, protection of openings
   with, 227
Openings, targeting, 220, 225–226,
   227, 231–232, 237
Partial Armor, 220
Plate, defenses offered by, 220,
   232–236
Rear Attacks, 227
Shield, use with, 94, 95

INDEX

Sidearms, use against, 38, 68, 175
Slices, protection from, 223–224, 227
Thrusts, protection from, 223, 224–225, 228–231, 238–239
Vs. Cutting Polearms, 214, 224, 225, 228, 232–234
Vs. Dagger, 238, 240
Vs. Spear, 199–200
Axe, One-Handed, 171–172, 174, 176

Bastard Sword: *see* Longsword
Bidenhänder: *see* Greatsword
Bill: *see* Polearms, Cutting
Bolognese, x, 10, 12, 13, 14
    Armor, 238,
    Cutting Polearms, 203, 205, 207, 208, 209, 211, 212, 213, 214, 215, 216, 217
    Dagger, 132, 138, 139, 142, 143, 144
    Greatsword, 148, 149, 150, 151, 152, 153, 154, 155, 156, 157, 215, 216, 217, 250
    Spear, 179, 180, 181, 182, 185, 186, 187, 188, 189, 190, 192, 193, 195, 196, 197, 198, 199, 200, 201, 251
    Sword Alone, 46, 47, 48, 97, 98, 129
    Sword and Buckler, 41, 45, 46, 47, 48, 50, 51, 52, 54, 55, 56, 58, 59, 60, 67, 97, 98, 130
    Sword and Cloak, 99, 100, 101, 104, 105, 106, 107
    Sword and Dagger, 108, 109, 110, 111, 112, 113, 115, 144
    Sword and Shield, 72, 77, 78, 79, 80, 84, 85, 86, 87, 89, 98
    Two Swords, 115, 116, 119, 120, 129
    Warfare, 246, 247, 250, 251, 252, 254
Buckler, Sword and, 10, 11, 41–69
    Buckler Punch, 56–59, 111, 145
    Buckler Shape, 44, 54–56
    Compared to Shield, 44, 52
    Construction, 44
    Definition, 41, 44, 72
    Grappling, 49, 58–59
    Grip, 50

Hand Protection, 49
Momentum, 44, 64
Multiple Opponents, use against, 62–67
Popularity, 45
Preferred Targets, 68
Reach, 62, 68
Targa, 54–56, 60
Use, active, 52
Use, passive, 49–52
Vs. Longsword, 67–69
Vs. Two Swords, 126–130
Weight (buckler), 44
Weight (sword), 41, 45
Wrist Cuts, use of, 44, 45, 50, 51, 64

Chainmail: *see* Armor
Cinquedea: *see* Dagger, Sword and
Cloak, Sword and, 10, 11, 97–108
    Availability, 98
    Casting, 105–108
    Compared to Other Companion Arms, 108
    Distance of Fight, 104, 106, 107–108
    How to Wear, 99
    Parry with, 99–101, 106
    Pushing with, 104
    Throwing: *see* Casting
    Time Period, 97–98
    Variety, 100
    Vulnerability, 101
Club: *see* Mace, *see* Warhammer
Cuts, 21, 22, 23, 24, 26, 29, 39, 44, 79–80, 101, 150, 206, 210, 223, 227

Dagger, 10, 11, 131–146
    Ambush, 141
    Cloak, use with, 11, 132
    Definition, 132
    Duels with, 143
    Escalation of Violence, 142
    Grappling, 133, 136, 138–140, 145
    Opponent's Dagger, focus on, 133–134, 143
    Parrying, difficulty of, 137, 138, 142–145
    Size, 132, 136

Speed, 134
Wounds, effect of, 144
Vs. Armor, 238, 240
Vs. Sword, 144–146
Dagger, Sword and, 10, 11, 108–115
   Attacks with, 111–112, 114–115
   Coordination Needed, 109
   Definition, 108
   Enchaining, 113–114
   Parrying with, 109–110
   Similarity to Shield, 108–109
   Stance, 109
Dall'Agocchie, Giovanni, 10
   Sword Alone, 97, 98, 129
   Sword and Buckler, 46, 47, 97, 98
   Sword and Cloak, 99, 101, 104
   Sword and Dagger, 108, 109, 112, 113
   Warfare, 247, 251, 252
Dardi, Filippo, 10, 12
Dei Liberi, Fiore, 1, 11, 12, 13, 14
   Armor, 219, 225, 231, 232
   Cutting Polearms, 204, 210, 212
   Dagger, 131, 132, 133, 134, 135, 136, 138, 140, 144
   Longsword, 17, 19, 20, 22, 26, 27, 29, 31, 32, 34, 37
   Mace, 173, 174
   Spear, 181, 182, 185, 191, 192, 194
   Warfare, 245, 246, 248
Destreza, La Verdadera: *see* La Verdadera Destreza
Di Grassi, Giacomo, 3, 5
Döbringer, Hans, 34
Dual Wielding: *see* Two Swords
Duels, 29–30, 47–48, 72, 79, 87–88, 105, 129, 131, 132, 142

Escrima Comun, x, 11, 12, 14
   Dagger, 132, 141, 142, 145, 146
   Greatsword, 94, 147, 148, 149, 150, 151, 152, 158, 159, 160, 161, 162, 163, 164, 165, 166, 167, 168, 169, 170
   Sword and Buckler, 45, 46, 52, 53, 54, 62, 64, 65, 66, 67, 127
   Sword and Cloak, 98, 99, 100, 101, 104, 106, 108
   Sword and Dagger, 97, 98, 108, 109, 110, 111, 112, 113, 114, 115, 120
   Sword and Shield, 71, 72, 76, 77, 78, 81, 82, 84, 93, 94
   Two Swords, 98, 116, 120, 121, 122, 123, 124, 125, 126
   Warfare, 246, 247
Experimental Archaeology, 7–8, 14, 75, 166

Juden, Andres, 34
Jud, Ott, 10

Forms, 46, 47, 56, 67, 149
Furlano de'i Liberi, Fiore: *see* dei Liberi, Fiore

Gambeson: *see* Armor
Ginetta: *see* Spear
Gladiatoria, 29, 195, 234
Glefen: *see* Spear
Godinho, Domingo Luis, 11, 12
   Dagger, 141, 142, 145, 146
   Greatsword, 94, 147, 148, 149, 150, 151, 152, 158, 159, 160, 161, 162, 163, 164, 165, 166, 167, 168, 169, 170
   Sword and Buckler, 52, 53, 62, 64, 65, 66, 67, 127
   Sword and Cloak, 100, 101, 104, 108
   Sword and Dagger, 97, 110, 111, 112, 113, 114, 120
   Sword and Shield, 71, 72, 76, 77, 78, 81, 82, 93, 94
   Two Swords, 121, 122, 123, 124, 125, 126
   Warfare, 246, 247
Greatsword, 11, 146, 147–170
   Bodyguard Weapon, use as, 161–162
   Definition, 147–149
   Grappling, 153–157
   Hilts, defensive features, 148, 154–155
   Impact, 166–167
   Maneuverability, 151, 152, 154, 160, 166–170

266   INDEX

Multiple Opponents, use against, 148, 158–167, 169–170
Pommel Strikes, 157
Reach, 94, 151, 153, 157, 158, 164, 168
Spins, 163–165
Time Period, 148
Unpredictability, 162, 165
Vs. Pike, 165–166
Vs. Polearm, 215–217
Vs. Shield, 167–170

Halberd: *see* Polearms, Cutting
Hand-and-a-Half Sword: *see* Longsword
HEMA, 3, 7–10, 17, 18, 19, 27, 30, 33, 38, 39, 46, 55, 56, 68, 79, 101, 103, 116–118, 127, 131, 132, 138, 149, 162, 178, 182, 183, 185, 186, 193, 194, 204, 216, 222, 244, 252–253
Historical European Martial Arts: *see* HEMA
Honor, 47–48, 129, 143–144, 157
Huntzfelt, Martin, 10
    Armor, 241, 242
    Dagger, 134, 137, 138

I.33: *see* Walpurgis Fechtbuch
Imbracciatura: *see* Shield, Sword and

Kal, Paulus, 10
    Cutting Polearms, 204, 212, 213
    Mace, 88, 92, 172
    Sword and Buckler, 44, 51, 59
    Sword and Shield, 91, 172
KDF, 10, 14,
    Armor, 225, 229, 230, 236, 238, 240, 241, 242
    Cutting Polearms, 204, 205, 206, 212, 213, 214
    Dagger, 132, 133, 134, 135, 137, 138, 139
    Longsword, 17, 18, 19, 20, 21, 22, 25, 26, 32, 33, 34
    Mace, 88, 92, 172

Spear, 182, 183, 184, 185, 187, 190, 193, 194, 225
Sword and Buckler, 44, 45, 46, 49, 50, 51, 53, 54, 55, 56, 59, 61, 62, 63, 64, 65
Sword and Shield, 72, 74, 88, 91, 92, 172
Warfare, 243, 245, 248
Kunst des Fechtens: *see* KDF

La Verdadera Destreza, 11
Lance: *see* Spear
Lancia: *see* Spear
Langes Messer: *see* Messer
Lanza: *see* Spear
Leather Armor: *see* Armor
Lecküchner, Johannes, 10
Lew, Jud, 33
Liechtenauer, Johannes, 10, 13, 14
    Armor, 238
    Longsword, 19, 22, 25
    Spear, 182, 183, 185, 225
    Warfare, 243, 245
Lignitzer, Andre, 10
    Dagger, 133, 134, 135, 139
    Sword and Buckler, 49, 50, 53, 54, 55, 61
Linen Armor: *see* Armor
Longshield: *see* Shield, Sword and
Longsword, 10, 11, 17–40, 188, 225, 229, 235
    Back edge, use of, 23–24
    Definition, 17–18, 20
    Grappling, use with, 30–33, 38, 145
    Leverage, 21, 38, 68, 94
    Maneuverability, 25, 39, 45, 68, 94–95
    Multiple opponents, use against, 33–38
    Pedagogical benefits, 19
    Pommel strikes, 26–29, 31, 39, 57, 235–236
    Pommel throwing, 29, 234–235
    Popularity, 19
    Reach, 20, 34, 36, 38, 40, 68, 94
    Thrown weapons, use against, 34

Vs. spear, 38–40
Vs. sword and buckler, 67–69
Vs. sword and shield, 93–95
Weight, 18, 45

Mace, 88, 171–178
    Comparison to Warhammer, 172, 174
    Shield, use with, 172
    Throwing, 172, 173–174
Mail: *see* Armor
Manciolino, Antonio, 10, 12, 69
    Cutting Polearms, 203, 207, 208, 212
    Dagger, 142, 144,
    Greatsword, 148, 149
    Spear, 179, 181, 182, 185, 186, 188, 189, 190, 192, 195, 199, 251
    Sword Alone, 46, 47, 48, 98
    Sword and Buckler, 46, 50, 51, 54, 56, 58, 59
    Sword and Dagger, 108, 109, 112
    Sword and Shield, 72, 80, 98
    Two Swords, 115, 116
    Warfare, 246, 251
Marozzo, Achille, 6, 10
    Cutting Polearms, 205, 209, 211, 215, 216, 217
    Dagger, 132, 138, 139, 142, 143
    Greatsword, 148, 149, 150, 151, 152, 153, 154, 155, 156, 157, 215, 216, 217, 250
    Spear, 181, 182, 185, 186, 187, 188, 192, 196, 197, 198, 199, 200, 201, 251
    Sword and Buckler, 41, 46, 54, 55, 56, 60, 130
    Sword and Cloak, 105, 106, 107
    Sword and Dagger, 108, 144
    Sword and Shield, 72, 78, 84, 85, 86, 87
    Two Swords, 115, 119
    Warfare, 246, 250, 251, 254
Messer, 10, 48–49
Meyer, Joachim, 10
Montante: *see* Greatsword
Monte, Pietro, 11

Armor, 221, 222, 227, 235
Cutting Polearms, 206, 211
Dagger, 137
Greatsword, 150
Spear, 185, 196,198, 199, 200
Sword and Buckler, 45, 51, 54
Sword and Shield, 72
Warhammer, 174, 175, 176, 177

Nissen, Jost von der, 34
Nuremberg Hausbuch, 7, 10, 14
    Longsword, 34

Pacheco de Narvaez, Luis, 11, 98, 115
Padded Armor: *see* Armor
Parrying, 21, 25, 50, 51, 52, 55, 68, 76–77, 78, 80, 99–101, 109–112, 116–118, 142, 145, 166–167, 177, 191, 197, 204–205, 211, 236
Partigiana: *see* Spear
Partisan: *see* Spear
Partisana: *see* Spear
Paurñfeyndt, Andre, 18
Picha: *see* Spear
Pike: *see* Spear
Plate Armor: *see* Armor
Polearms, Cutting, 10, 11, 203–217
    Butt, use of, 204, 205, 206, 208, 212, 217
    Cuts with, 210–212
    Grappling, 212–214
    Finesse, 205, 206, 208, 217
    Force, 203, 204
    Reach, 94
    Speed, 206, 211, 212
    Thrusts with, 207–208, 211, 212
    Terminology, 204, 209–210
    Vs. Armor, 214, 224, 225, 228, 232–234
    Vs. Greatsword, 215–217
Poleaxe: *see* Polearms, Cutting
Pons de Perpiñan, Jaime, 11, 98
Preußen, Nicklass, 34

Román, Francisco, 11
Ringeck, Sigmund, 5, 10

268 INDEX

Armor, 225,
    Longsword, 25, 26
Ronca: see Polearms, Cutting
Rotella: see Shield, Sword and

Schiltschlac: see Buckler, Buckler Punch
Self-Defense, 33, 44, 48, 56, 62–63, 99, 108, 120–126, 131, 141, 144, 160, 161
    Bystanders, importance of, 123–124
Seydenfaden, Hans, 10
Sharpness, 101–103, 116–118, 154
Shield, Sword and, 10, 11, 46, 71–95
    Definition, 71–72
    Grips, 72–75
    Hooking, with and against, 73–74, 87, 91, 94
    Imbracciatura, 72, 85–87
    Longshield, 10, 72, 83, 87
    Preferred Targets, 79–81, 85, 94
    Range of Motion, 75–78
    Reach, 87–88, 94
    Rotation, 73–74, 89–90, 91–92
    Shield Punch, 73, 75, 83–84, 86–87, 95
    Spikes, 85, 87, 89, 91
    Straps, use of, 84
    Use, active, 76–77, 79, 81
    Use, passive, 77–79
    Vs. Greatsword, 167–170
    Vs. Longsword, 93–95
    Vs. Spear, 200–202
Sidesword, definition, 42, 43
Slices, 33, 105, 154, 208, 223, 224, 227
Spadone: see Greatsword
Spear, 10, 11, 179–202
    Blows, percussive, 183, 187, 190, 193
    Butt, use of, 194
    Footwork, 192–193, 196
    Grappling, 40
    Lance, 180, 183–184
    Maneuverability, 39, 188, 189, 197
    Pike, 180, 181, 192, 195
    Power, 183
    Preferred Targets, 189–190, 197, 198, 201
    Reach, 38, 40, 94
    Shield, and, 84, 196–200
    Social Status, 180–182
    Speed, 183, 189
    Sword, transition to, 194–195, 201
    Throwing, 183–186, 197
    Throwing, *rotella*, 199
    Variety, 180
    Vs. Armor, 199–200
    Vs. Longsword, 38–40
    Vs. Greatsword, 215–217
    Vs. Sword and Shield, 200–202
Spiedo: see Spear
Sport Fencing, 6, 13, 20
Storta: see Messer
Sword Alone, 10, 11, 46–49, 98, 129
Sword Terminology, 18–19, 42–43, 147–148

Talhoffer, Hans, 10, 103
    Armor, 229
    Cutting Polearms, 205, 206
    Sword and Buckler, 44, 49, 54, 56, 62, 63, 64, 65
    Sword and Shield, 74, 88, 92
    Warfare, 248
Targa: see Buckler, Sword and
Targe, 54
Thrusts, 21, 22, 23, 39, 79, 150, 160, 183, 199, 207–208, 211, 223, 228–231, 238
Torre, Pedro de la, 11, 98, 115, 148
Two-Handed Sword: see Greatsword, Definition
Two Swords, 11, 115–130
    Attacks with, 119
    Compared to Sword and Dagger, 116, 119
    Coordination Required, 116
    Difficulty, 115, 128
    Jumps and Spins, 122, 126
    Multiple Opponents, use against, 120–126

Social Context, 128–130
Swords Used, 115
Time Period, 115
Use in Real Combat, 115
Vs. Sword and Buckler, 126–130

Vadi, Philippo, 11
Viggiani, Angelo, 12, 13
Von Danzig, Peter, 10
Spear, 184, 185
Von Danzig, Pseudo-Peter:
Armor, 225, 229, 230, 236, 238, 240
Longsword, 22, 32
Spear, 187, 194

Walpurgis Fechtbuch, 13, 45, 49, 61, 178
Warfare, 87, 160, 165–166, 170, 207, 243–255
Fencing, military style, 251–255
Military Fencing, 244
Treatises, addressed by, 245–251
Warhammer, 54, 171–178
Comparison to Mace, 172, 174
Comparison to Sword, 176–177
Hand Protection, 54, 176–177
Hooking, 175–176
Two Hands, use in, 174–175

Zettel: *see* Liechtenauer, Johannes
Zweihänder: *see* Greatsword

www.ingramcontent.com/pod-product-compliance
Lightning Source LLC
Chambersburg PA
CBHW071335080526
44587CB00017B/2847